THE ECONOMICS OF NON-CONVEX ECOSYSTEMS

THE ECONOMICS OF NON-MARKET GOODS AND RESOURCES
VOLUME 4

Series Editor: Dr. Ian J. Bateman

Dr. Ian J. Bateman is Professor of Environmental Economics at the School of Environmental Sciences, University of East Anglia (UEA) and directs the research theme Innovation in Decision Support (Tools and Methods) within the Programme on Environmental Decision Making (PEDM) at the Centre for Social and Economic Research on the Global Environment (CSERGE), UEA. The PEDM is funded by the UK Economic and Social Research Council. Professor Bateman is also a member of the Centre for the Economic and Behavioural Analysis of Risk and Decision (CEBARD) at UEA and Executive Editor of Environmental and Resource Economics, an international journal published in cooperation with the European Association of Environmental and Resource Economists (EAERE).

Aims and Scope

The volumes which comprise The Economics of Non-Market Goods and Resources series have been specially commissioned to bring a new perspective to the greatest economic challenge facing society in the 21st Century; the successful incorporation of non-market goods within economic decision making. Only by addressing the complexity of the underlying issues raised by such a task can society hope to redirect global economies onto paths of sustainable development. To this end the series combines and contrasts perspectives from environmental, ecological and resource economics and contains a variety of volumes which will appeal to students, researchers, and decision makers at a range of expertise levels. The series will initially address two themes, the first examining the ways in which economists assess the value of non-market goods, the second looking at approaches to the sustainable use and management of such goods. These will be supplemented with further texts examining the fundamental theoretical and applied problems raised by public good decision making.

For further information about the series and how to order, please visit our Website
http://www.wkap.nl/series.htm/ENGO

The Economics of Non-Convex Ecosystems

Edited by

Partha Dasgupta
Cambridge University, Faculty of Economics, Cambridge, UK

and

Karl-Göran Mäler
*Beijer International Institute of Ecological Economics,
Royal Swedish Academy of Sciences, Stockholm, Sweden*

KLUWER ACADEMIC PUBLISHERS
DORDRECHT / BOSTON / LONDON

A C.I.P. Catalogue record for this book is available from the Library of Congress.

ISBN 1-4020-1864-9 (PB)
ISBN 1-4020-1945-9 (HB)

Published by Kluwer Academic Publishers,
P.O. Box 17, 3300 AA Dordrecht, The Netherlands.

Sold and distributed in North, Central and South America
by Kluwer Academic Publishers,
101 Philip Drive, Norwell, MA 02061, U.S.A.

In all other countries, sold and distributed
by Kluwer Academic Publishers,
P.O. Box 322, 3300 AH Dordrecht, The Netherlands.

Printed on acid-free paper

Reprinted from *Environmental & Resource Economics,* Vol. 26: 499 – 685 (2003).

All Rights Reserved
© 2004 Kluwer Academic Publishers
No part of this work may be reproduced, stored in a retrieval system, or transmitted
in any form or by any means, electronic, mechanical, photocopying, microÞlming, recording
or otherwise, without written permission from the Publisher, with the exception
of any material supplied speciÞcally for the purpose of being entered
and executed on a computer system, for exclusive use by the purchaser of the work.

Printed in the Netherlands.

Contents

The Economics of Non-Convex Ecosystems: Introduction
P. Dasgupta, K.-G. Mäler   1

Scale and Scaling in Ecological and Economic Systems
J. Chave, S. Levin   29

Convex Relationships in Ecosystems Containing Mixtures
of Trees and Grass
R.J. Scholes   61

Managing Systems with Non-convex Positive Feedback
W.A. Brock, D. Starrett   77

The Economics of Shallow Lakes
K.-G. Mäler, A. Xepapadeas, A. de Zeeuw   105

Multiple Species Boreal Forests – What Faustmann
Missed
A.-S. Crépin   127

Evaluating Projects and Assessing Sustainable Development in Imperfect Economies
K.J. Arrow, P. Dasgupta, K.-G. Mäler   149

The Economics of Non-Convex Ecosystems: Introduction

PARTHA DASGUPTA[1] and KARL-GÖRAN MÄLER[2]
[1]*Cambridge University, Faculty of Economics, Sidgwick Avenue, Cambridge CB3 9DD, UK (E-mail: partha.dasgupta@econ.cam.ac.uk);* [2]*Beijer International Institute of Ecological Economics, Stockholm, Sweden*

Abstract. The word "convexity" is ubiquitous in economics, but absent from economics. In this paper we explain why, and show what difference it makes to economic analysis if ecosystem non-convexities are taken seriously. A simple proof is provided of the connection between "self-similarity" and "power laws". We also provide an introduction to each of the papers in the Symposium and draw out the way in which they form a linked set of contributions.

Key words: bifurcation points, environmental Kuznets curve, hysteresis, irreversibility, non-convexity, Pontryagin Principle, power laws, property rights, separatrix, structural stability, thresholds

1. Economists' Convexities and Nature's Non-Convexities

The word "convexity" is ubiquitous in economics, but absent from ecology. There is a reason for each. As prices are prominent in modern transactions, it is but natural that we would wish to uncover the ways in which the price system is capable of functioning as a resource allocation mechanism. In recent years economists have identi ed not only the way in which prices aggregate dispersed pieces of information, but also the sense in which they re ect the relative scarcities of goods and services. We now know that the price system can be an ef cient allocation mechanism if transformation possibilities among goods and services – in and over time – constitute a convex set.[1] However, except in the case of partial economic systems, as in models of industrial organization (Tirole 1993), or of systems harbouring very speci c forms of non-convexities, as in modern growth models (Jones 1998) or in models of poverty traps based on the connection between nutritional status and human productivity (Dasgupta and Ray 1986, 1987; Dasgupta 1997) or in models of spatial economies (Fugita, Krugman and Venables 1999), we still do not have a clear understanding of the mechanisms by which resources are allocated in non-convex environments. So we economists continue to rely on the convexity assumption, always hoping that it is not an embarrassing simpli cation.

Ecologists have no comparable need to explore the structure of convex sets. They are interested in identifying pathways by which the constituents of an ecosystem interact with one another and with the external environment. A large body of empirical work has revealed that those pathways in many cases involve transformation possibilities among environmental goods and services that, together, constitute non-convex sets. Often the non-convexities reflect positive feedbacks in Human-Nature interactions. Mathematical ecologists therefore study the structural stability of ecosystems and the sizes and shapes of their basins of attraction for given sets of environmental parameters.[2] Such notions as the "resilience" of ecosystems are expressions of this research interest.[3]

The price mechanism is especially problematic in economic systems characterised by positive feedback processes. We now know that in such environments it may prove impossible to decentralise an efficient allocation of resources by means exclusively of prices. Efficient mechanisms would typically involve additional social contrivances, such as (Pigouvian) taxes and subsidies, quantity controls, social norms of behaviour, and so forth. This was proved formally in a justly famous article by Starrett (1972), who showed that for certain types of non-convexities associated with environmental pollution, a competitive price equilibrium simply does not exist: markets for pollution would be unable to equate demands to supplies. Starrett's non-convexities are present when losses traceable to environmental pollution are bounded. If the market price for pollution in such situations were negative (i.e., the polluter has to pay the pollutee), pollutees' demand would be unbounded, while supply would be bounded. On the other hand, if the price were non-negative, demand would be zero, while supply, presumably, would be positive.[4]

Starrett's finding implied that private property rights to environmental pollution would not be capable of sustaining an efficient allocation of resources by means of the price system. In a subsequent note, Starrett (1973) demonstrated by means of an example that if property rights are awarded to polluters, even such a non-price resource allocation mechanism as the core may not exist. But he showed (Starrett 1972) that a suitably chosen set of pollution taxes, together with a system of competitive markets for other goods and services (assuming that the latter constitute a convex sector), would be capable of supporting an efficient allocation of resources. As there are no markets for pollution in such an allocation mechanism, the problem of equating supply to demand in pollution activities is bypassed. The moral would appear to be that social difficulties arising from the non-convexities can be overcome if the State were to assign property rights in a suitable way – permitting private rights to the convex sector, but reserving for itself the right to control emissions and discharges.[5]

2. Ecological Thresholds and the Environmental Kuznets Curve

Despite the ecologist's strictures, we economists have remained ambivalent toward Nature's non-convexities. Often enough, that ambivalence reveals itself only indirectly. For example, it is even today commonly thought that, to quote an editorial in the UK's The Independent (4 December 1999), "... (economic) growth is good for the environment because countries need to put poverty behind them in order to care"; or, to quote The Economist (4 December 1999, p. 17), "... trade improves the environment, because it raises incomes, and the richer people are, the more willing they are to devote resources to cleaning up their living space".

The view's origin can be traced to World Bank (1992), which observed an empirical relationship between GNP per head and atmospheric concentrations of industrial pollutants. Based on the historical experience of OECD countries, the authors of the document suggested that, when GNP per head is low, concentrations of such pollutants as the sulphur oxides increase as GNP per head increases, but that when GNP per head is high, concentrations decrease as GNP per head increases further.[6] Among economists this relationship has been christened the "environmental Kuznets curve".[7] The moral that would appear to have been drawn from the nding is that resource degradation is reversible: degrade all you want now, Earth can be relied upon to rejuvenate it later if you require it.

The presumption is false. Nature's non-convexities are frequently a manifestation of positive feedback processes, which in turn often means the presence of ecological thresholds. But if a large damage were to be in icted on an ecosystem whose ability to function is conditional on it being above some threshold level (in size, composition, or whatever), the consequence would be irreversible.[8] The environmental Kuznets curve was detected for mobile pollutants. Mobility means that, so long as emissions decline, the stock at the site of the emissions declines. However, reversal is the last thing that would spring to mind should a grassland ip to become covered by shrubs, or should the Atlantic gulf stream shift direction or come to a halt, or should a source of water disappear, or should an ocean shery become a dead zone owing to over shing. As a metaphor for the possibilities of substituting manufactured and human capital for natural capital, the relationship embodied in the environmental Kuznets curve has to be rejected.[9]

Although non-convexities are prevalent in global ecosystems (e.g., ocean circulation, global climate), it is as well to emphasise the spatial character of many positive feedback processes. The latter have a direct bearing on the rural poor in the world's poorest regions. Eutrophication of ponds, or salinization of soil, or biodiversity loss in a forest patch involve crossing ecological thresholds at a spatially localised level. Similarly, the metabolic pathways between an individual's nutritional status and his or her capacity to work, and those between a person's nutritional and disease status involve positive feedback.[10] Studies of extreme poverty based on aggregation at the regional or national level can therefore mislead greatly. The spatial con nement of many of the non-convexities inherent in Human-Nature interactions needs always to be kept in mind.

The connection between rural poverty in the world's poorest regions and the state of the local ecosystems should be self-evident. When wetlands, inland and coastal sheries, woodlands, forests, ponds and lakes, and grazing elds are damaged (owing, say, to agricultural encroachment, or urban extensions, or the construction of large dams, or organizational failure at the village level), traditional dwellers suffer. For them – and they are among the poorest in society – there are frequently no alternative source of livelihood. In contrast, for rich eco-tourists or importers of primary products, there is something else, often somewhere else, which means that there are alternatives. Whether or not there are substitutes for a particular resource is therefore not only a technological matter, nor a mere matter of consumer taste: among poor people location can matter too. The poorest of the poor experience non-convexities in a way the rich do not. Even the range between a need and a luxury is context-ridden. Macroeconomic reasoning glosses over the heterogeneity of Earth's resources and the diverse uses to which they are put – by people residing at the site and by those elsewhere.[11]

3. Motivation Behind this Symposium

Despite their prevalence, we economists continue to show little interest in Nature's non-convexities. For example, the now enormous literature on "green accounting" has been built largely on the backs of economic models in which non-convexities are safely out of sight.[12] In view of this, the Beijer International Institute of Ecological Economics in Stockholm invited a group of ecologists and economists to study the economics of non-convex ecosystems. Beginning in 1998, the group met regularly.[13] We knew that much fundamental work had already been accomplished on non-linear dynamical systems. Some of it had already been used to study the economics of non-convex social environments.[14] Our idea therefore was to use these techniques so as to study proto-typical Human-Nature interactions. Speci c applications were very much in our mind.

As our understanding of the issues improved, we felt it would be desirable too if the articles we prepared were written expansively. So we encouraged ourselves also to review what has already been achieved in the subject. We felt that the collection could then serve as a self-contained body of work, useable in graduate courses in environmental and resource economics, perhaps useful as well to researchers who wish to work in this eld. In order to increase accessibility to the collection, we felt it would be appropriate to publish it in a journal, rather than as a book. We are therefore most grateful to the members of the Editorial Board of Environmental and Resource Economics, especially Ian Bateman and Kerry Turner, for the enthusiasm they have shown toward our enterprise from the time we approached them, and for the encouragement they have given us to produce the collection in the form that it now appears.

4. Macro Regeneration Functions of Ecosystems

4.1. PRODUCTION FUNCTIONS AND COMMODITY POSSIBILITY SETS

In contrast to the production scale-economies studied in traditional price theory, the non-convexities associated with ecological thresholds manifest themselves across time. Nevertheless, there is a formal connection between the two types of non-convexities.

To see this, consider the economist's old stand-by, a one-commodity, constant population economy, where the commodity in question is durable and non-deteriorating. Time is continuous and the economy is deterministic.[15] Let K_t (≥ 0) and C_t (≥ 0) be the "capital" stock and the flow of consumption, respectively, at t (≥ 0). Output is assumed to be given by the production function $F(K)$, where F is differentiable everywhere, $F(0) = 0$ and $F(K) > 0$ for some $K > 0$. An extreme assumption, much used in general equilibrium theory, is that output is *freely disposable*. Subsequently we will see what the assumption involves and why its violation is significant for the subject of this Symposium.[16]

Let I_t denote investment at t. Since output is freely disposable, we may express the balance of flows at each date in the economy by the inequality,

$$I_t \leq F(K_t) - C_t, \quad \text{for } t \geq 0, \text{ and } K_0 \;(> 0) \text{ is given,} \tag{1a}$$

which, without rigorous justification, we write as,

$$dK_t/dt \leq F(K_t) - C_t, \quad \text{for } t \geq 0, \text{ and } K_0 \;(> 0) \text{ is given.} \tag{1b}$$

A *programme* is a complete forecast of the economy, from the present (t = 0) to infinity. A programme can be expressed as $\{K_t, C_t, dK_t/dt\}_0^\infty$. For simplicity of exposition, we suppose that any programme satisfying (1b) is feasible; which is to say that the economy is not subject to any other constraint.

The question is whether the set of feasible programmes is convex.

The answer depends on whether F is *concave everywhere* (henceforth, *concave*). To confirm this, assume that F is concave. Consider any two feasible programmes, which we write as $\{K_t^*, C_t^*, dK_t^*/dt\}_0^\infty$ and $\{K_t', C_t', dK_t'/dt\}_0^\infty$. By definition, both satisfy (1b). Hence,

$$dK_t^*/dt \leq F(K_t^*) - C_t^*, \quad \text{for } t \geq 0, \text{ and } K_0^* = K_0, \tag{2}$$

$$\text{and} \quad dK_t'/dt \leq F(K_t') - C_t', \quad \text{for } t \geq 0, \text{ and } K_0' = K_0. \tag{3}$$

Now choose a number γ, where $0 \leq \gamma \leq 1$. Define

$$\bar{C}_t = \gamma C_t^* + (1-\gamma)C_t' \text{ and } \bar{K}_t = \gamma K_t^* + (1-\gamma)K_t'. \tag{4}$$

We wish to confirm that $\{\bar{K}_t, \bar{C}_t, d\bar{K}_t/dt\}_0^\infty$ also satisfies (1b).

From (2) and (3), we have,

$$\gamma dK_t^*/dt \leq \gamma F(K_t^*) - \gamma C_t^*, \quad \text{for } t \geq 0, \text{ and } K_0^* = K_0, \tag{5}$$

$$\text{and } (1-\gamma)dK_t'/dt \leq (1-\gamma)F(K_t') - (1-\gamma)C_t', \quad \text{for } t \geq 0, \text{ and } K_0' = K_0. \tag{6}$$

Since F is concave,

$$F(\bar{K}_t) = F(\gamma K_t^* + (1-\gamma)K_t') \geq \gamma F(K_t^*) + (1-\gamma)F(K_t'). \tag{7}$$

From (2)–(7) we conclude that

$$d\bar{K}_t/dt \leq F(\bar{K}_t) - \bar{C}_t, \qquad \text{for } t \geq 0, \text{ and } \bar{K}_0 = K_0, \tag{8}$$

which means that the set of programmes satisfying (1b) is convex.

If F(K) is not concave, the above argument does not work, because inequality (7) cannot be guaranteed. In fact, if F(K) is not concave, it is possible to find a pair of programmes that satisfy inequality (1b) and a convex combination of the two that does not satisfy inequality (1b). *This means that if the function F(K) is not concave, the set of feasible programmes is not convex.*

4.2. WELFARE ECONOMICS IN A CONVEX WORLD

Let us review welfare economics in a world where F(K) is concave. Both consumption and the capital stock are assumed to be "goods" (i.e., neither is a pollutant), and we imagine that the flow of social welfare at t is a strictly concave function, $U(C_t, K_t)$. U is assumed to be twice partially differentiable everywhere. Let δ (> 0) be the utility rate of discount. Intertemporal welfare is taken to be of the form,

$$\int_0^\infty U(C_t, K_t) \exp(-\delta t) dt, \qquad \text{where } \delta > 0. \tag{9}$$

The standard problem in welfare economics is to locate that programme which maximizes (9) subject to (1b). But since output is a good and there are no bad byproducts, it would be silly, ever, to dispose of any output, which is another way of saying that the shadow price of output is positive. So we may express the optimization problem as being one of locating that programme which maximizes (9) subject to:

$$dK_t/dt = F(K_t) - C_t, \qquad \text{for } t \geq 0, \text{ and } K_0 \text{ (> 0) given.} \tag{10}$$

As is now hugely familiar, the problem lends itself to analysis by the techniques developed by Pontryagin. In view of the assumptions that have been made about F and U, we know that the Hamiltonian of the maximization problem is a concave function of the state and control variables. This means that the optimum can be implemented with the help of intertemporal accounting prices in a decentralized economic environment, the point to which we drew attention in Section 1 of the Introduction.[17]

The literature on optimum economic growth has shown that if U depends only on C, the Pontryagin conditions yield a unique stationary point (a saddle point), the target that ought to be aimed at no matter what is the value of K_0. To the best of our knowledge, Kurz (1968) was the first to note that if U depends not only on C, but also on K (as is the case of the U postulated here), the Pontryagin optimality conditions may possess multiple stationary points, meaning that in the space of K and its co-state variable, there are multiple basins of attraction.[18] Kurz also showed that the long run features of the optimum programme depend on K_0. In other words, history matters. We conclude that even when the Hamiltonian of an optimization problem is concave in the state and the control variables, the Pontryagin optimality conditions can have multiple stationary points.

In the economics of forestry it has been common to assume that the growth function of a tree's biomass is concave. Denoting the biomass of a tree by K, it is assumed that K's growth function is quadratic. If trees are not harvested, biomass accumulates as,

$$dK_t/dt = F(K_t) = aK_t - bK_t^2, \qquad a, b > 0, \text{ and } K_0 > 0. \qquad (11)$$

Equation (11) possesses two stationary points: $K = 0$ and $K = a/b$. The former is unstable, while the latter is stable. The system therefore possesses a single basin of attraction. Starting with a seedling, a tree's biomass grows as a logistic function. This gives rise to the famous Faustmann Rule for the age at which a synchronized forest should be felled for its timber and then reseeded immediately. Underlying Faustmann's Rule is the assumption that a forest is valuable only for its timber, which means that tree biomass has no direct worth (say, as a habitat for biodiversity). As is well known, the biomass felled according to the Rule, at regular intervals, is less than a/2b (the point at which F(K) is maximum). Given that F in equation (11) is concave, the Faustmann Rule can in principle be implemented by a price system.[19]

4.3. CONVEX-CONCAVE F

The famous Bretherton-Holt regeneration function for fish biomass takes the form of equation (11). However, that F(K) may not be concave was realized many years ago by fisheries experts. Denoting the biomass of a single-species fishery by K, it is commonly assumed in fisheries economics that F(K) is *convex-concave*, that is, *F(K) is convex at low values of K, but concave beyond some value of K*. A rigorous analysis of the optimal management of a convex-concave resource was, however, not provided until Skiba (1978). As we shall see presently, several of the contributions in this Symposium are economic analysis of ecosystems whose natural regeneration functions (F(K) in our notation) are convex-concave.

The problem is that, as F is convex-concave, feasible programmes do not constitute a convex set even if catch is freely disposable. The price system is therefore generally not viable for implementing the optimum programme, the point with which we began this Introduction.

5. Micro Foundations of Emergent Macro Properties

Thus far we have reviewed the macro properties of some very special ecosystems. In a wide ranging discussion, Chave and Levin (this issue) study certain emergent properties of systems that are governed by processes operating at different scales of activity.[20] Exploring both spatial and temporal scales, the authors show how natural scientists have tried to provide the micro foundations of the macro properties of natural adaptive systems. Chave and Levin offer a rich set of examples, drawn from ecology, geology, and condensed matter physics, to illustrate common traits among adaptive systems. Of particular interest are systems that are *scale invariant* (or "self similar"). The authors note that they display *power laws*. The relationship between metabolic rate and body mass is one example of a power law, the allometric dependency of species number on area is another. As the proof that scale invariant systems display power laws is simple and intuitive, we offer it here for completeness.

Suppose that two scalar quantities, x and y (both non-negative), are related by a scale invariant function,

$$y = f(x). \quad (12)$$

In order to change scales, define,

$$z = \alpha y, \text{ and } w = \beta x, \quad \text{where } \alpha, \beta > 0. \quad (13)$$

Since f is scale invariant, it must be that

$$z = Af(w), \quad \text{where } A (= A(\alpha, \beta)) > 0. \quad (14)$$

But equations (12)–(14) can hold together if and only if f is homogeneous, that is,

$$f(x) = Bx^\gamma, \quad \text{where } B > 0 \text{ and } \gamma \text{ is a constant.} \quad (15)$$

(15) is the implied power function.

As in economics, it is easier to study the macro dynamics of an ecosystem directly without peering at micro foundations. In the four papers that follow the Chave-Levin study, the authors investigate the macro dynamics of three types of ecosystems: savannahs, shallow lakes, and boreal forests. For tractability, the ecosystems are described in terms of differential equations involving only a few state variables. Depending on the context, the state variables are interpreted as resource stocks or as environmental qualities. Humanity is taken to be the user of the ecosystem. The models are deterministic.

6. Mixed Tree-Grass Ecosystems

Scholes (this issue) explains why the functional relationship between tree cover and grass cover in the African savannahs is convex for much of its range. Starting at zero, if tree cover were to increase, grass cover would decline, but at a rate that would define a convex function (at least initially), meaning that the set of grass-tree configurations circumscribed by the curve is non-convex. Scholes argues that because the economic value of the savannahs lies mainly in grass products, and because the main mechanism (fires) controlling the balance between grass and trees is dependent on grass, the convex functional relationship between trees and grass results in the simplest cases in two stable savannah configurations: thin tree cover and dense tree cover. In such cases the savannahs possess two basins of attraction. A landscape of thin tree cover is able to support grazing, but one that has dense tree cover is not able to do so.

The existence of multiple basins of attraction in non-convex systems is a recurrent theme of this Symposium. While Scholes offers an explanation for the observed landscapes of the savannahs (expressed in terms of trees and grass and the ecological pathways mediating their mix), the remaining contributions study Human-Nature interactions in terms of the way Nature is (or, alternatively, should be) managed by Human communities. In each of the systems studied here, every basin of attraction is attracted to an *equilibrium point* of the system.[21]

7. Human Intervention in Non-Convex Ecosystems

Human intervention in an ecosystem takes the form of resource extraction or pollutant discharge. In positive as opposed to normative analysis the intervention is taken to be given, its consequences to the ecosystem are then studied. In theoretical work the intervention is frequently taken to be constant over time, which means that it can be regarded as a parameter of the ecosystem. We illustrate this by considering, in turn, resource extraction and pollution discharge in the context of two simple models.

7.1. CONVEX-CONCAVE GROWTH FUNCTIONS: RESOURCE EXTRACTION

Dasgupta (1982, ch. 6) constructed a model of an open access fishery. Denoting fish biomass by K, he assumed that the fishery's growth function, $F(K)$, is

$$F(K) = -a + bK - cK^2, \quad a, b, c, (b^2 - 4ac) > 0, \quad \text{if } K > 0, \text{ and} \quad (16)$$
$$F(K) = 0 \quad \text{if } K = 0.$$

$F(K)$, so defined, is convex-concave. (To confirm this, observe that $F(K)$ is discontinuous at $K = 0$.) Notice also that $F(K) = 0$ at three values: 0, $K_1 = [b - (b^2 - 4ac)^{1/2}]/2c$, and $K_2 = [b + (b^2 - 4ac)^{1/2}]/2c$. K_1 is a *threshold*: if biomass were to fall below it, the fishery would die.

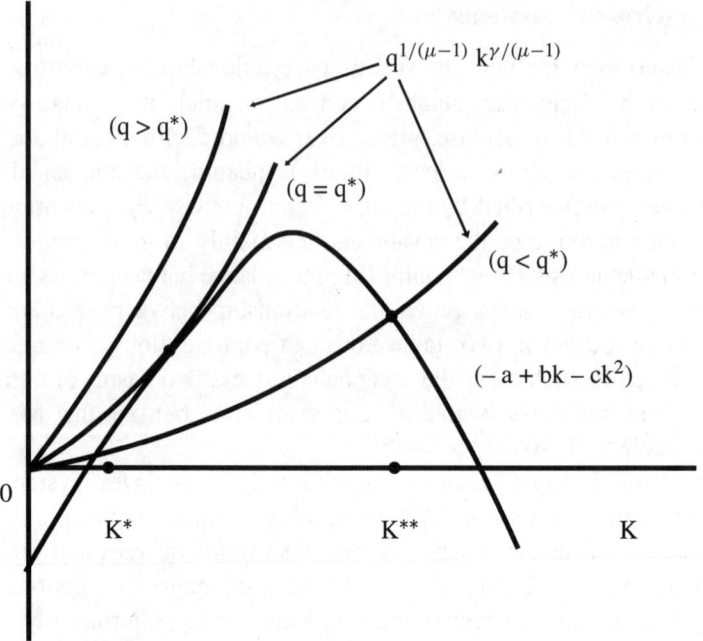

Figure 1. Open access fishery.

If C is catch, the cost of harvest is taken to be $\beta K^{-\gamma} C^{\mu}$, where $\mu > 1$ and β, $\gamma > 0$. The idea here is that, while unit cost of harvest is an increasing function of harvest (there is crowding among fishermen), harvest cost is a decreasing function of stock (search costs decline if fish biomass increases).

Free entry (and exit) into the fishery is assumed to be an instantaneous adjustment process, resulting in zero profit at all times. This means that if p is the market price of a unit of fish biomass,

$$p = \beta K_t^{-\gamma} C_t^{(\mu-1)}. \tag{17}$$

Writing $q \equiv p/\beta$ and using (16) and (17), the net growth rate of fish biomass (for $K_t > 0$) is

$$dK_t/dt = (-a + bK_t - cK_t^2) - q^{1/(\mu-1)} K_t^{\gamma/(\mu-1)}. \tag{18}$$

a, b, c, μ, γ, and q are parameters of the dynamical system. Since q reflects the fish market and an aspect of fishing technology, we shall vary q so as to study the qualitative properties of equation (18).

Figure 1 is based on the right hand side of equation (18). We have drawn the figure for the case where $\gamma > (\mu - 1)$. Observe that there is a critical value of q, call it q*, such if q = q*, the curves $(-a + bK - cK^2)$ and $q^{1/(\mu-1)} K^{\gamma/(\mu-1)}$ are tangent to each other.

If q > q*, equation (18), allied to the second part of (16), has a unique stationary point: K = 0. This means that the fishery has a single basin of attraction and is doomed: there is overfishing to such an extent that, in finite time, the fishery becomes extinct. This is the case advanced by Hardin (1968), who claimed as a universal rule that freedom in the commons spells ruin. But the thesis is correct only if, relative to the harvesting technology, the resource is valuable ($q > q^*$). Even under open access, matters are different when $q < q^*$.

To confirm this, notice that if $q < q^*$, the fishery has three stationary points, namely, $K = 0$ and the two stationary solutions of equation (18), which we have denoted as K^* and K^{**} in Figure 1. It is simple to infer from the figure that $K = 0$ and $K = K^{**}$ are (locally) stable, while K^* is unstable. K^* is thus a *separatrix*: it separates the two basins of attraction, each of which contains a stationary point ($K = 0$ and $K = K^{**}$, respectively). We conclude that if $K_0 > K^*$, the fishery survives despite free entry. Hardin's analysis turns out to have been wrong for this case: if the value of fish relative to harvesting costs is low, there is not enough entry to pose a danger to the fishery. Because there is a qualitative break in the dynamical properties of the fishery under free entry at $q = q^*$, q^* is called a *bifurcation* point of the system.

7.2. CONVEX-CONCAVE POLLUTION RECYCLING FUNCTIONS

7.2.1. *The model*

We now study a pollution problem that has been much analysed in recent years: phosphorus discharge into a shallow, fresh water lake. We use the occasion also to develop a number of additional notions.[22]

Consider a system whose dynamics are described by equation (10). However, so as to make contact with the papers by Brock and Starrett and by Mäler, Xepapadeas, and de Zeeuw in this Symposium, we now interpret K_t (≥ 0) to be the stock of a pollutant in an ecosystem and C_t (≥ 0) the pollution discharge into the system. K can be taken to reflect the state of the ecosystem.

It would be absurd to assume that pollutants can be disposed of freely. Kneese, Ayers, and d'Arge (1972) noted that mass has to be conserved, so that what is produced by society must eventually find a place of residence, in one form or another. This they formalized by requiring that economic activities must satisfy a "materials-balance condition".[23] In the present case, equation (10) is an equation (unlike equation (1b), not a weak inequality), because pollution cannot be disposed of freely.

Assume that $C_t = C$, a constant. C is the human intervention. The pollution's dynamics are therefore given by the equation,

$$dK_t/dt = F(K_t) + C, \quad \text{for } t \geq 0, \text{ and } K_0 \ (> 0) \text{ given.} \tag{19}$$

F(K) represents the *net* natural regeneration rate of the pollutant. One of the services ecosystems provide us with is the breaking down of pollutants. Organic pollutants are decomposed by microbes, atmospheric carbon dioxide gets absorbed by the oceans, and so forth. Such natural forms of decay are included in F. But there are cases (as with phosphorus in shallow lakes; see below) where the pollutant has a limited ability to grow as well. So, in what follows we assume that (i) F is differentiable everywhere, (ii) F(0) = 0, (iii) F(K) > 0 for some K, and (iv) F(K) < 0 for all K sufficiently large.

The term *resilience* refers to the stability of ecosystems. It has been used alternatively to denote the size of the basin of attraction in which it currently resides and the speed of convergence back to its current state following a perturbation to it. In order to explore such problems, it pays first to study the dynamics of the ecosystem when C = 0. In this case,

$$dK_t/dt = F(K_t), \qquad \text{for } t \geq 0, \text{ and } K_0 \ (> 0) \text{ given.} \tag{20}$$

K = 0 is a stationary point of the system. How many other stationary points does the system possess?

If F is concave, there is precisely one other stationary point, and it is stable, while K = 0 is unstable. But if F is not concave, equation (20) can have any number of stationary points. In view of the conditions we have imposed on F, the function cannot be convex everywhere. So we look for *convex-concave* forms.

Consider the simplest convex-concave form, namely,

$$F(K) = bK^2/(1 + K^2) - \lambda K, \qquad \text{where } b, \lambda > 0. \tag{21}$$

The positive feedback in the regeneration process governing K is given by the first term on the right hand side (RHS) of equation (21), which, as can be readily confirmed, is convex-concave, with an upper bound of b. The second term on the RHS of (21) is the rate at which the ecosystem is cleansed of the pollutant. Combining (19) and (21) we have

$$dK_t/dt = C + bK_t^2/(1 + K_t^2) - \lambda K_t, \qquad K_0 \ (> 0) \text{ given.}[24] \tag{22}$$

Equation (22) contains three parameters: C, b, and λ. We wish to know how the ecosystem's character depends on them. One expects that mostly the global properties of the ecosystem would vary continuously with the parameters. One should also expect that there are manifolds partitioning the parameter space into regions, such that the ecosystem's structure is the same at every point in any given region, but differs from the structure in the region adjacent to it. Such manifolds are said to "bifurcate" the system's properties. To study the bifurcations, we take b and λ to be given and we vary C. The reason we permit C to vary is that C denotes human intervention and we could in principle control it.

So consider the equation

$$bK^2/(1 + K^2) = \lambda K. \tag{23}$$

Real solutions of equation (23) are the stationary points of equation (22) with C = 0.

We begin by assuming that $\lambda/b > 1/2$, which is to say that the pollutant decays rapidly. In this case (23) has only one real solution: it is K = 0. Simple graphics

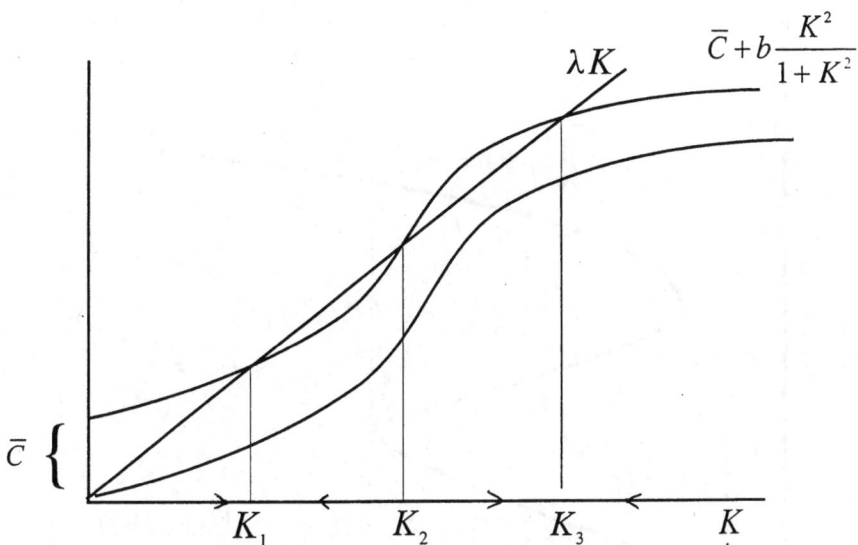

Figure 2. Pollution equilibria: the reversible case.

(Figure 2) confirm, however, that there are values of C for which equation (22) has three (real) stationary points. Assuming one such value, we label the stationary points as K_1 (<) K_2 (<) K_3, respectively. K_2 is unstable, while K_1 and K_3 are locally stable. K_2 is the *separatrix* of the system – the point that separates the two basins of attraction of the ecosystem.

7.2.2. *Ecosystem flips*

Continuing to hold b and λ constant, let us now reduce C. It is simple to confirm visually that the unstable stationary point (continue to label it K_2) and the larger of the two locally stable stationary points (continue to label it K_3) get closer to each other continuously. It is simple to confirm as well that there is a critical value of C, call it C^*, for which K_2 and K_3 coincide to form a point that is stable from the right, but unstable from the left. C^* is a *bifurcation point* of the system: if $C < C^*$, the ecosystem possesses a unique (stable) stationary point, whereas if $C > C^*$ (but $C < C^{**}$; see below), it possesses three stationary points. In short, the system's structure changes discontinuously at C^*.[25]

In contrast, suppose C were to increase. It is simple to confirm visually (Figure 2) that the unstable stationary point (continue to label it K_2) and the smaller of the two locally stable stationary points (continue to label it K_1) would get closer to each other continuously, until, at a critical value of C, call it C^{**}, the two would coincide, to form a point that is unstable from the right, but stable from the left. C^{**} is another bifurcation point of the system: if $C > C^{**}$, the ecosystem possesses a unique (stable) stationary point, whereas if $C < C^{**}$ (but $C > C^*$), it possesses three stationary points.

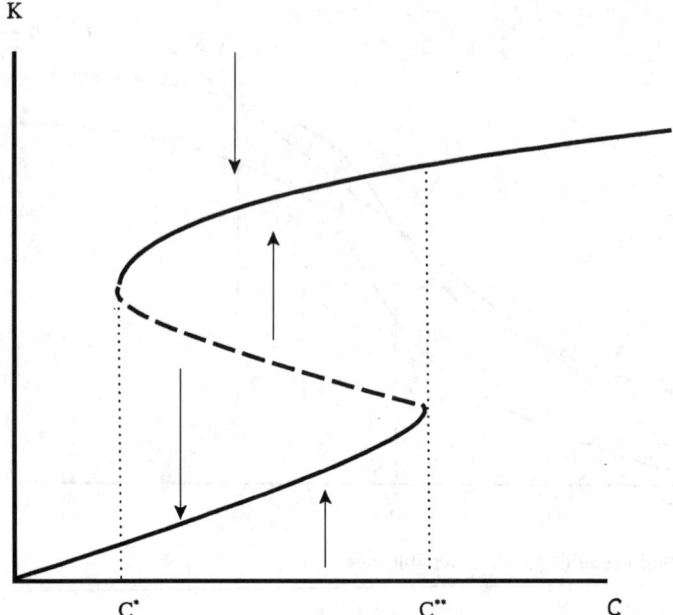

Figure 3. Ecosystem dynamics for constant pollutant inflow: the reversible case.

In Figure 3 we have drawn the equilibrium values of K as a correspondence of C for a given pair of values of b and λ. Equilibrium K is unique when $C < C^*$. For C in the interval $[C^*, C^{**}]$, the curve depicting K as a correspondence of C bends back and then back again, to reflect the fact that equation (22) possesses three stationary points. The two upward sloping portions of the correspondence consist of (locally) stable stationary values of K, whereas the downward sloping portion consists of unstable stationary points.

We now conduct a thought experiment. Begin in a situation where $C < C^*$. We know that equilibrium K is small. We would like to discover how the system would change if C were to increase in a predictable way. Rather than try to integrate equation (22), we simplify by imagining that C increases slowly relative to the speed of adjustment of K_t. By "slowly" we mean that at each C the ecosystem is able to equilibrate itself. If C were to increase under such conditions, K would increase continuously along the lower arm of the curve until $C = C^{**}$, at which point equilibrium K would "flip" to the upper arm of the curve. The ecosystem therefore undergoes a discrete change at C^{**}. Further increases in C would lead to a continual increase in K along the upper arm of the curve in Figure 3.

Ecosystem flips have been observed many times and at many scales. Shallow lakes have been known to flip from clear to turbid water in a matter of months, village tanks in a matter of weeks, garden ponds in a matter of hours. Insect populations have been known to crash or explode in a matter of days. Larger ecosystems generally take longer to flip at their bifurcation points because the underlying processes operate over greater distances, and are therefore slower. Grasslands in sub-Saharan Africa can take more than a decade to change into shrublands. The "salt conveyor" that drives global ocean circulation would probably take between decades and a century to shut down (or change direction) if the Greenland ice cover were to melt at rates estimated in current models of global warming.[26] The fossil records suggest that the interglacials and glacials of ice ages have appeared only occasionally, but have arrived and departed "precipitously" – the flips occuring over several thousand years.[27] And so on.

7.2.3. *Hysteresis in ecosystem dynamics*

Now suppose we are interested in reversing the process. Start with $C > C^{**}$ and reduce it slowly. Figure 3 shows that on the return journey, K declines continuously along the upper arm so long as $C > C^*$. This means that for C in the interval $[C^*, C^{**}]$, K remains higher than it was on the onward journey. To put it another way, the ecosystem displays *hysteresis*. However, at $C = C^*$ the ecosystem "flips" to the lower arm of the curve in Figure 3. Further declines in K would occur continuously if C were reduced further. We conclude that, even though the ecosystem displays hysteresis, environmental degradation is *reversible*: given enough time, K can be made to be as small as we like if C were reduced sufficiently. This is the intellectual basis of the environmental Kuznets curve, mentioned earlier. As we have just confirmed, it would be a correct view of future possibilities if the pollutant's decay rate were sufficiently large.

7.2.4. *Irreversibility in ecosystem damage*

But now consider a less happy possibility. Suppose that $\lambda/b < 1/2$, which is to say that the pollutant decays slowly. In this case equation (23) possesses three real solutions. One is $K = 0$, while the other two are positive. Figure 4, which is the counterpart of Figure 2, depicts this case. We now use Figure 4 to construct Figure 5, which plots the equilibrium values of K as a correspondence of C. In contrast to Figure 3, the curve bends backward to cut the vertical axis.

Let us conduct the thought experiment again. Suppose we begin in a situation where both C and K are low, which means that the system is on the lower arm of the curve in Figure 5. As C increases, K increases continuously until the bifurcation point $\hat{C}$ is reached. At this point the ecosystem flips to a higher value of K. However, once that happens, the system is incapable of reversing itself. Declines in C would certainly reduce K, but as Figure 5 shows, even if C were reduced to zero, the system would remain on the upper arm of the curve, at a higher value of K than it did to begin with. Not only does the ecosystem suffer from hysteresis, but environmental degradation is now in addition *irreversible*: the system is unable to return to where it had been in the beginning.

We are used to the intuitive idea that the presence of thresholds in ecosystems means that large damages to them are irreversible. The above analysis has

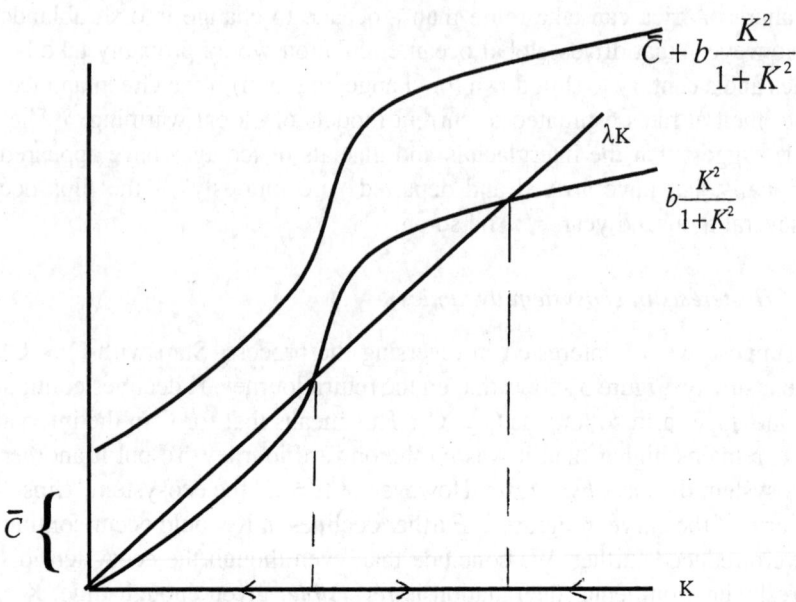

Figure 4. Pollution equilibria: the irreversible case.

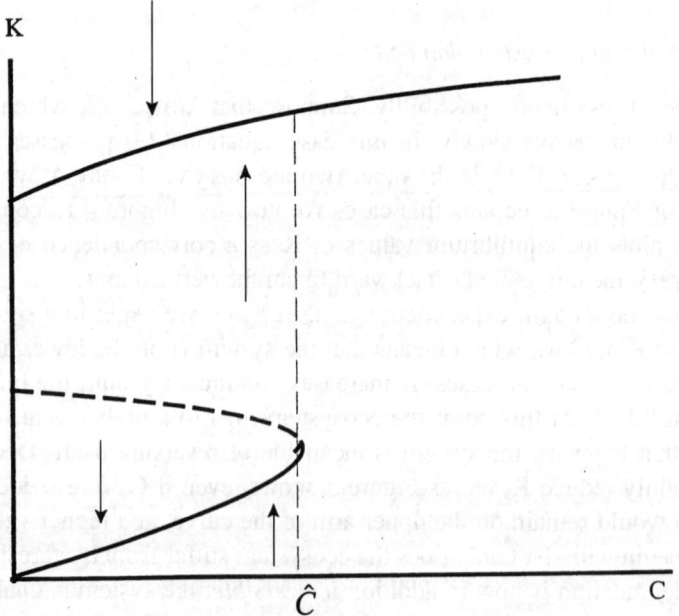

Figure 5. Ecosystem dynamics for constant pollutant inflow: the irreversible case.

shown that large damages can be irreversible even if an ecosystem does not have thresholds. Positive feedback, allied to a slow decay rate for the pollutant in the ecosystem, is all that is needed.

8. Optimum Intervention

Particular interest attaches to optimum intervention. Brock and Starrett interpret K_t in equation (22) to be phosphorus load in the water column of a shallow lake. Phosphorus is taken to be a byproduct of agricultural production in the neighbourhood of the lake. As a flow, it is therefore a "good". However, as a stock in the lake, it is a "bad". The flow of utility is taken to be,

$$U(C_t, K_t) = \log C_t - hK_t^2, \quad \text{where } h > 0, \qquad (24)$$

and intertemporal welfare is assumed to be of the form expressed in (9), which for convenience we rewrite here:

$$\int_0^\infty U(C_t, K_t) \exp(-\delta t) dt, \quad \text{where } \delta > 0. \qquad (25)$$

The dynamical system can be expressed as

$$dK_t/dt = C_t + bK_t^2/(1 + K_t^2) - \lambda K_t, \quad K_0 \, (>0) \text{ given}. \qquad (26)$$

The idea is to maximise (25) subject to equation (26), given that U satisfies (24).

Brock and Starrett show that, for a range of values of δ, h, λ, and b, the pair of equations (involving the state and the (Pontryagin) co-state variables) that optimum programmes must necessarily possess three stationary points. Two (call them K_1 and K_3, with $K_1 < K_3$, corresponding to what could be interpreted to be the oligotrophic and eutrophic state, respectively) are saddle points, while the intermediate point (call it K_2) is a spiral source (i.e., it is unstable).[28] They note that, for any K_0, the two trajectories that asymptote to K_1 and K_3, respectively, satisfy not only the necessary conditions for optimality, but the transversality condition as well. This would suggest that, in order to judge which is the correct trajectory, one would need to compute expression (25) for each trajectory and then compare the two values – a tiresome requirement. However, the authors show that there is a value of phosphorus stock, K^*, such that if $K_0 > K^*$, the optimum programme asymptotes to K_3; but that if $K_0 < K^*$, it asymptotes to K_1. In other words, history matters.[29] The also confirm that if by fluke $K_0 = K^*$, the two trajectories that asymptote to K_1 and K_3, respectively, are equally desirable. The authors refer to K^* as a Skiba point, in recognition of Skiba (1978), who uncovered the existence of such points of indifference.

Locating K^* is no easy matter. The disturbing conclusion is that if the planner were by mistake to think that K_0 is greater than K^*, the path he would choose be wrong: the planner would regard the eutrophic state to be the optimal target, whereas he ought instead to aim for the oligotrophic state. It would seem that, in order to determine the optimum programme, there is no alternative to having to compute the value of expression (25) along both candidate trajectories.

Brock and Starrett also show that there is a range of parameter values for which the optimality conditions possess a single stationary point (a saddle point), so that there are no Skiba points. The set of parameter values that separate the two regions in the space of parameter values defines a bifurcation surface.[30]

Mäler, Xepapadeas, and de Zeeuw study a case that parallels the analysis offered in Section 7.1 (Dasgupta 1982). There are a number of non-cooperative farmers discharging phosphorus into a lake. The pollutant in the lake is a public bad, but each farmer's discharge is a private good to that farmer. Assuming that equilibrium behaviour yields an open loop solution for the differential game played by the farmers, the authors identify cases where the farmers drive the lake to a eutrophic state, even though the optimal policy would have been to aim toward an oligotrophic state. As noted above, once the eutrophic state is reached, it may be impossible to bring the lake back into an oligotrophic state. More generally, if environmental problems are neglected over a long period, even with all the good will in the world it may prove impossible to reverse the situation subsequently.

9. Multi-Species Forests

The defining proposition in the economic theory of forestry is Faustman's Rule, which identifies the optimum age at which trees should be harvested in a single species forest. Earlier we observed that for deriving Faustmann's Rule, the rate at which a synchronised forest's biomass grows is assumed to be a concave, quadratic function of that biomass (equation (11)). In multi-species forests, however, the regeneration rate of a species' biomass depends not only on its own biomass, but also on the biomass of other species. For example, when plant species compete for sunlight and nutrients, the biomass of each species has a negative effect on the others' regeneration rates.[31] One conclusion follows immediately: Because of species interdependence, Faustmann's Rule does not hold.

Forest species are not all plants. Crepin (this issue) studies a model of boreal forests, consisting of pine, birch, and moose. Moose feed on both pine and birch, so both tree species have a positive effect on moose biomass. In contrast, moose biomass has a negative effect on the growth of pine and birch. Given the biomasses of pine and birch (moose), the regeneration rate of moose (birch) biomass is assumed to be a concave, quadratic function. In contrast, Crepin argues that when pine biomass is small, young pines establish better with increasing biomass. So she assumes that, for any pair of values of moose and birch biomasses, the regeneration rate of pine is a convex-concave function of its own biomass. Each of the three species is assumed to have an economic worth at each date, both as harvest and as stock. The social objective function is the present discounted value of the flow of economic worth, from the present to the indefinite future.

Technically, Crepin's problem is considerably more complicated than the single state-variable model of the shallow lake. Nevertheless, she is able to report substantial results. She shows that forests may possess multiple basins of attraction, meaning that the optimum harvesting policy in a multi-species forest can depend on the state at which it is found to begin with. The author derives optimum stationary state management rules and characterises the nature of the steady states. Faustmann's Rule holds for neither birch nor pine.

10. Welfare Economics in Imperfect Economies

Even in non-convex environments, optimum management policies are known to satisfy Pontryagin's Maximum Principle. Brock-Starrett and Crepin apply the principle to derive the conditions that optimum programmes must necessarily satisfy. Brock and Starrett emphasize that, where the Pontryagin necessary conditions for optimality possess Skiba points, adding transversality conditions would not suffice. What is required in such situations is the sheer brute force of computing social welfare along the candidate programmes and then comparing them.

The Brock-Starrett-Crepin formulations are in the spirit of modern welfare economic theory, where policy analysis is conducted in the context of a State (or regulatory authority) that selects those policies that maximize social welfare subject to technological, environmental, and institutional constraints. In modern welfare economics, what differentiates the "first-best" from the plethora of "second-best" policies it is possible to envisage is the inclusion of institutional constraints when deriving the latter. The State is assumed to act on behalf of its citizens even in second-best economies.

One can argue that this is a misleading approach to policy analysis in dysfunctional societies. It is hard to imagine the sense in which governments in what are demonstrably failed states may be said to be optimizing on behalf of their citizens. However, it may not be absurd to imagine that even in the most corrupt and predatory of governments there are honest people. It can be safely assumed that such figures are minor officials, involved in making marginal decisions (a road here, a local environmental protection plan there, and so on). What language does welfare economics have to speak to such people in what we may euphemistically call *imperfect economies*?

In an earlier work, Dasgupta (2001b) extended intertemporal welfare economics to imperfect economies by determining rules that can be used to evaluate small perturbations to macroeconomic forecasts. A forecast is based on a reading of technological and environmental possibilities, and on the behaviour of households, firms, communities, and the State. A perturbation to the forecast is interpreted as a project (or more generally, a small policy change) under the jurisdiction of the honest civil servant. Dasgupta showed that the required evaluation rule involves the use of accounting prices that can in principle be estimated by perturbing the forecast. The rule itself is to check whether the present discounted value of the flow of social profits generated by the perturbation is positive. In other words, the choice criterion is the one that has for long been advocated for social cost-benefit analysis in optimizing economies. Dasgupta's analysis did not require the economy to be convex.

The entire intellectual machinery deployed by Dasgupta (2001b) was taken from Dasgupta and Mäler (2000), who had developed a criterion for assessing whether or not intertemporal social welfare is *sustained* along an economic forecast. The authors had shown in the context of a simple economic model with constant population that the same accounting prices as those that can be used in social cost-benefit analysis can also be used to compute an index for assessing welfare sustainability. The index in question was shown to be a comprehensive measure of *wealth*, or in other words, the social value of the entire stock of the economy's capital assets, inclusive of manufactured, human, and natural capital assets.[32] The Dasgupta-Mäler analysis did not require the economy to be convex.

Arrow, Dasgupta, and Mäler (this issue) extend the findings in the earlier two papers by answering three related questions in the context of a very general model: (1) How should accounting prices be estimated? (2) How should policy changes in an imperfect economy be evaluated? (3) How can one check whether intertemporal social welfare will be sustained along a projected economic programme? The authors do not presume that the economy is convex. They show that the same set of accounting prices should be used both for policy evaluation and for assessing whether or not intergenerational social welfare along a given economic path will be sustained. They also confirm that a comprehensive measure of wealth, computed in terms of the accounting prices, can be used as an index for problems (2) and (3) above. They extend the earlier results by including changes in the size of population. They also show how uncertainty in future possibilities can be accommodated in the comprehensive measure of wealth. The bulk of their paper contains rules for estimating the accounting prices of several specific environmental natural resources, transacted in a few well known economic institutions.

11. Time and Space

The papers included here on the welfare economics of environmental resources take time seriously, but not space. In this they reflect the current state of environmental and resource economics accurately. And yet, the spatial dimension is of the utmost importance. At the grandest scale, terrestrial ecosystems differ from marine ecosystems. But even within each there is a wide distribution, covering height, depth, and spread. Individual members of a population not only interact with one another (involving both inter- and intra-species exchanges), but they affect and are in turn affected by the abiotic processes at different sites. The evolution of landscapes is modulated by such interactions. In any given patch, populations breed and die. They also disperse to other sites within and outside the patch. The physical, chemical, and biological processes at work operate at different scales, both spatially and temporally. Together they give rise to spatially structured populations and landscapes. What we observe is a system that is not only spatially and temporally patchy, but also modular.[33]

Despite the self-evident importance of the spatial aspect of Nature, welfare economic theory is not yet spatially sensitive. If a community has to extract resources from its local ecosystem, which patches should it harvest, what should it harvest, and when? Applied studies inform us of the harvesting strategies rural communities in the African savannahs, say, have adopted over time in response to the spatial and temporal character of their local ecosystems. Frequently enough, rationale behind their strategies have been unearthed as well. But are they optimal from the community's point of view? How would we know? As matters stand, even the mathematical tools that would be required to address those problems are not familiar in the economics literature. It is a potentially very fruitful area for future research.

12. Good and Bad Positive Feedbacks, and Rich and Poor People

Contemporary models of economic growth are dismissive of the importance of Nature. In their extreme form, growth models contain an assumed positive link between the creation of ideas (technological progress) and population growth in a world where the natural-resource base comprises a fixed, indestructible factor of production.[34] The models do involve positive feedback, but of a Panglossian kind.

The error in this literature lies in the fact that Nature is not fixed and industructible, but rather, it consists of degradable resources (agricultural soil, watersheds, fisheries, and sources of fresh water; more generally, ecological services). It may be sensible to make the wrong assumption for studying a period when natural-resource constraints did not bite, but it is not sensible when studying development possibilities today. The latter move is especially suspect when no grounds are offered by growth theorists for supposing that technological progress can be depended upon indefinitely to more than substitute for an ever-increasing loss of the natural-resource base.

In any event, we should be sceptical of a theory that places such enormous burden on an experience not much more than two hundred years old. Extrapolation into the past is a sobering exercise: over the long haul of history (a 5,000 years stretch, say, upto about two hundred years ago), economic growth even in the currently-rich countries was for most of the time not much above zero.[35]

Positive feedback in ecological and metabolic pathways are reasons why the prospects of economic betterment among the world's poorest are much harder than among the rich. For one thing, the poor in a poor economy have to operate on the boundary of the non-convex region of their nutrition-productivity possibilities, whereas people who possess sufficient assets are able to get onto the boundary of their convex region.[36] For another, the non-convexities the poor face can also be a reflection of their inability to obtain substitutes for depleted natural resources. As we noted earlier, resource depletion for the poor can be like crossing a threshold: their room for maneouver is circumscribed hugely once they cross. In contrast, the rich can usually "substitute" their way out of problems.

The simultaneous presence of two types of positive feedback – one enabling many to move up in their living standard, the other keeping many others in poverty – may explain the large scale persistence of absolute poverty in a world that has been growing wealthier on average by substituting manufactured and human capital for natural capital. For human well-being, policies matter, as do institutions, but the local ecology matters too. If in this Introduction we have focussed on the positive feedback mechanisms that operate at the downside of life, it is because degradation of the natural-resource base is felt first by the poor, not the rich.

Acknowledgement

We are most grateful to Mukul Majumdar for helpful comments.

Notes

1. Koopmans (1957) and Debreu (1959) are the classic expositions of this. For completeness, here is the definition of convexity *of a set*:
   A commodity vector, say z, is a *convex combination* of commodity vectors x and y if z is a weighted average of x and y, where the weights are non-negative and sum to unity (that is, $z = \alpha x + (1-\alpha)y$ for some $\alpha \varepsilon \, [0,1]$). A set of commodity vectors is said to be *convex* if every convex combination of every pair of commodity vectors in the set is in the set. A set is *non-convex* if it is not convex.
2. May (1975) and Murray (1993) contain illuminating accounts of both the mathematical and ecological reasoning.
3. See Perrings et al. (1995), Levin and Barrett et al. (1998), and Gunderson and Holling (2002). We discuss these issues formally in Section 7.
4. In an earlier classic, Arrow (1971) had observed that markets for externalities would suffer from another problem: no matter whether the externalities are positive or negative, the markets would be "thin".
5. Over the past three decades much work has been done by economists to develop the theory of environmental policy. Baumol and Oates (1975) continues to be the outstanding treatise on the subject.
6. See also Cropper and Griffiths (1994) and Grossman and Krueger (1995).
7. It is, of course, a misnomer. The original Kuznets curve, which was an inverted U, related income inequality to real national income per head on the basis of historical cross-country evidence.
8. In Section 7 we review the mathematics of bifurcations and irreversibilities.
9. For further discussions of the environmental Kuznets curve, see Arrow and Bolin et al. (1995) and the responses it elicited in symposia built round the article in *Ecological Economics*, 1995, 15(1); *Ecological Applications*, 1996, 6(1); and *Environment and Development Economics*, 1996, 1(1). See also the special issue of *Environment and Development Economics*, 1997, 2(4); and Dasgupta, Levin and Lubchenco (2000).
10. WHO (1983, 1985) and Spurr (1988, 1990) are classic publications on the relationship between nutritional status and human productivity. For evidence on synergies between nutritional and disease status, see Scrimshaw et al. (1968) and Harrison and Waterlow (1990).

11. See the interchange between Johnson (2001) and Dasgupta (2001a) on this. For a more detailed analysis of the connection between environmental and resource economics and the economics of poverty, see Dasgupta (1982, 1993, 2000, 2003).
12. See for example, Lutz (1993), Heal (1998), and Weitzman (2000).
13. The meetings were financed by a generous grant from the MacArthur Foundation in Chicago.
14. Brock and Malliaris (1989) is an excellent treatise in that area.
15. As is well known, discrete- and continuous-time systems can differ fundamentally in regard to their stability properties. The decision to work in continuous time is therefore not an innocuous one. As is also well known, the structure of "non-linear" dynamical systems even in continuous time is currently understood best when the number of state variables does not exceed 2. Readers should note that the analysis that follows is undertaken on one-dimensional natural systems.

    Finally, it goes without saying that restricting the study to deterministic systems is hugely limiting. Barring Chave and Levin and the final section of the paper by Arrow, Dasgupta, and Mäler, all the articles in this issue study deterministic systems. For stability properties of random dynamical economies in discrete time, see Bhattacharya and Majumdar (2003).
16. Koopmans (1957) contains an especially thoughtful discussion of "free disposability".
17. The Pontryagin co-state variable associated with the optimum is the required accounting price. Kamien and Schwartz (1991) contains a good account of dynamic optimization theory. Accounting pirces are also called shadow prices.
18. See also Keeler, Spence and Zeckhauser (1972).
19. Johansson and Löfgren (1985) contains a thorough account.
20. May (1989) contains a fine, non-technical exposition of a number of issues raised by Chave and Levin.
21. For a fine study of the economics of chaotic systems, see Majumdar, Mitra and Nishimura (2000).
22. Each of the additional notions can also be developed by means of the fishery's model of the previous sub-section.
23. Mäler (1974) contains a general equilibrium analysis of resource allocations subject to the materials-balance condition.
24. Close variants of equation (22) have been postulated for a number of natural systems. Here are three examples:

    (1) In order to explain periodic infestations of the spruce budworm in boreal forests, Ludwig, Jones, and Holling (1978) postulated that the budworm's population, K_t, changes in accordance with the equation
    $$dK_t/dt = \alpha K_t - \beta K_t^2 - bK_t^2/(1 + K_t^2), \qquad (\alpha, \beta, b > 0), \tag{22a}$$
    where the final, forcing term denotes predation by birds.

    (2) The account of the Atlantic thermohaline circulation in Rahmstorf (1995, 2002) can be formalised in terms of an equation not dissimilar to (22). Temperature and salt gradients across the North and South Atlantic gives rise to the circulation. K_t is taken to be the North Atlantic deep water flow (travelling south) and C is the amount of fresh water entering, say, the surface of the North Atlantic (in part from ice melts). Circulation can come to a halt if C is too large.

    (3) Scheffer (1997) has used equation (22) to characterise the dynamics of phosphorus deposit in the water column of shallow lakes when C is the inflow of phosphorus into a lake from neighbouring farms. The papers by Brock and Starrett, and Mäler, Xepapadeas, and de Zeeuw (this issue) study that model.

    (4) Vegetation cover in the savannahs depends on rainfall, but rainfall in turn depends on vegetation cover. Denoting rainfall by C_t and vegetation (in biomass) by K_t, suppose, as a first approximation, that
    $$C_t = \alpha K_t \text{ and } dK_t/dt = bC_t^2/(1 + C_t^2) - \lambda K_t, \qquad (\alpha, b, \lambda > 0). \tag{22b}$$
    The pair of equations (22a, b) are variants of (22).
25. Mathematicians call this a "saddle-node bifurcation".
26. Rahmstorf (2002).

27. See van Andel (1994) and Rahmstorf (1995).
28. Although, for ease of exposition, we are using the same notation, the points K_1, K_2, and K_3 here are not the same as the points K_1, K_2, and K_3 in the previous section.
29. Skiba (1978) showed that in non-convex economies the optimality conditions may possess multiple stationary points even if the utility function is independent of stocks.
30. Wagener (2003) has shown that the bifurcation curves in the above system are what mathematicians call "heteroclinic bifurcations".
31. See for example, Pastor and Mladenoff (1992).
32. Pearce and Atkinson (1993) had earlier proved that an inclusive measure of wealth can be used to assess the sustainability of intertemporal welfare in an optimizing world with constant population.
33. Levin (1999) has an excellent non-technical account of this.
34. Kremer (1993) develops such a model to account for 1 million years of world economic history.
35. See Fogel (1994, 1999), Johnson (2000), and especially Maddison (2001). The claim holds even if the past two hundred years were to be included. The rough calculation is simple enough: World per capita output today is about 5,000 US dollars. The World Bank regards one dollar a day to be about as bad as it can be: people would not be able to survive on anything much less than that. It would then be reasonable to suppose that 2,000 years ago per capita income was not less than a dollar a day. So, let us assume that it was a dollar a day. This would mean that per capita income 2,000 years ago was about 350 dollars a year. Rounding off numbers, this means very roughly speaking that, per capita income has risen about 16 times since then. This in turn means that world income per head has doubled every 500 years, which in its turn means that the average annual rate of growth has been about 0.14 percent per year, a figure not much in excess of zero.
36. Dasgupta and Ray (1986). To illustrate, the undernourished are at a severe disadvantage in their ability to obtain food: the quality of work they are able to offer is inadequate for obtaining the food they require if they are to improve their nutritional status. Over time undernourishment can be both a cause and consequence of someone falling into a poverty trap. Because undernourishment displays hysteresis, such poverty can be dynastic: once a household falls into a poverty trap, it can prove especially hard for descendents to emerge out of it. Many poverty studies involving econometric exercises (including many that explore nutrition and health) assume linear relationships among the stipulated variables. By construction they are incapable of detecting the non-convexities inherent in metabolic and ecological processes.

References

Arrow, K. J. (1971), 'Political and Economic Evaluation of Social Effects of Externalities', in M. Intriligator, ed., *Frontiers of Quantitative Economics* **1**. Amsterdam: North Holland.

Arrow, K. J., B. Bolin, R. Costanza, P. Dasgupta, C. Folke, C. S. Holling, B.-O. Jansson, S. Levin, K.-G. Mäler, C. Perrings and D. Pimentel (1995), 'Economic Growth, Carrying Capacity, and the Environment', *Science* **268**(5210), 520–521.

Arrow, K. J., P. Dasgupta and K.-G. Mäler (2003), 'The Genuine Saving Criterion and the Value of Population', *Economic Theory* **21**(2), 217–225.

Baumol, W. M. and W. Oates (1975), *The Theory of Environmental Policy*. Englewood Cliffs, NJ: Prentice-Hall.

Bhattacharya, R. and M. Majumdar (2003), *Random Dynamical Systems*, Mimeo. New York: Department of Economics, Cornell University, Forthcoming, Cambridge University Press, 2004.

Brock, W. A. and A. G. Malliaris (1989), *Differential Equations, Stability and Chaos in Dynamic Economies*. Amsterdam: North Holland.

Brock, W. A. and D. A. Starrett (2003), 'Non-Convexities in Ecological Management Problems', *Environmental and Resource Economics* **26**, 575–602.
Chave, J. and S. Levin (2003), 'Scale and Scaling in Ecological and Economic Systems', *Environmental and Resource Economics* **26**, 527–557.
Cropper, M. L. and C. Griffiths (1994), 'The Interaction of Population Growth and Environmental Quality', *American Economic Review* **84**(Papers & Proceedings), 250–254.
Dasgupta, P. (1982), *The Control of Resources*. Cambridge, MA: Harvard University Press.
Dasgupta, P. (1993), *An Inquiry into Well-Being and Destitution*. Oxford: Clarendon Press.
Dasgupta, P. (1997), 'Nutritional Status, the Capacity for Work and Poverty Traps', *Journal of Econometrics* **77**(1), 5–38.
Dasgupta, P. (2000), 'Population, Resources, and Poverty: An Exploration of Reproductive and Environmental Externalities', *Population and Development Review* **26**(4), 643–689.
Dasgupta, P. (2001a), 'On Population and Resources: Reply', *Population and Development Review* **26**(4), 748–754.
Dasgupta, P. (2001b), *Human Well-Being and the Natural Environment*. Oxford: Oxford University Press.
Dasgupta, P. (2003), 'World Poverty: Causes and Pathways', in B. Pleskovic and N. H. Stern, eds., *Annual Bank Conference on Development Economics 2003*. Washington, DC: World Bank, Forthcoming, 2004.
Dasgupta, P., S. A. Levin and J. Lubchenco (2000), 'Economic Pathways to Ecological Sustainability', *BioScience* **50**(4), 339–345.
Dasgupta, P. and K.-G. Mäler (2000), 'Net National Product, Wealth, and Social Well-Being', *Environment and Development Economics* **5**(1), 69–93.
Dasgupta, P. and D. Ray (1986), 'Inequality as a Determinant of Malnutrition and Unemployment, 1: Theory', *Economic Journal* **96**(4), 1011–1034.
Dasgupta, P. and D. Ray (1987), 'Inequality as a Determinant of Malnutrition and Unemployment, 2: Policy', *Economic Journal* **97**(1), 177–188.
Debreu, G. (1959), *Theory of Value*. New York: John Wiley.
Fogel, R. W. (1994), 'Economic Growth, Population Theory, and Physiology: The Bearing of Long-Term Processes on the Making of Economic Policy', *American Economic Review* **84**(3), 369–395.
Fogel, R. W. (1999), 'Catching Up with the Economy', *American Economic Review* **89**(1), 1–19.
Fugita, M., P. Krugman and A. Venables (1999), *The Spatial Economy: Cities, Regions, and International Trade*. Cambridge, MA: MIT Press.
Grossman, G. M. and A. B. Krueger (1995), 'Economic Growth and the Environment', *Quarterly Journal of Economics* **110**(2), 353–377.
Gunderson, L. and C. S. Holling (2002), *Panarchy*. Washington, DC: Island Press.
Hardin, G. (1968), 'The Tragedy of the Commons', *Science* **162**, 1243–1248.
Harrison, G. A. and J. C. Waterlow, eds. (1990), *Diet and Disease in Traditional and Developing Countries*. Cambridge: Cambridge University Press.
Heal, G. M. (1998), *Valuing the Future: Economic Theory and Sustainability*. New York: Columbia University Press.
Johansson, P.-O. and K.-G. Löfgren (1985), *The Economics of Forestry and Natural Resources*. Oxford: Basil Blackwell.
Johnson, D. Gale (2000), 'Population, Food, and Knowledge', *American Economic Review* **90**(1), 1–14.
Johnson, D. Gale (2001), 'On Population and Resources: A Comment', *Population and Development Review* **27**(4), 739–747.
Jones, C. I. (1998), *Introduction to Economic Growth*. New York: W.W. Norton.
Kamien, M. I. and N. L. Schwartz (1991), *Dynamic Optimization*. Amsterdam: North Holland.

Keeler, E., A. M. Spence and R. Zeckhauser (1972), 'The Optimal Control of Pollution', *Journal of Economic Theory* **4**(1), 19–34.
Kneese, A. V., R. U. Ayers and R. C. d'Arge (1972), *Economics and the Environment: A Materials Balance Approach*. Washington, DC: Resources for the Future.
Koopmans, T. C. (1957), 'The Price System and the Allocation of Resources', in T. C. Koopmans, ed., *Three Essays on the State of Economic Science*. New York: McGraw Hill.
Kremer, M. (1993), 'Population Growth and Technological Change: One Million B.C. to 1990', *Quarterly Journal of Economics* **108**(3), 681–716.
Kurz, M. (1968), 'Optimal Economic Growth and Wealth Effects', *International Economic Review* **9**(3), 348–357.
Levin, S. (1999), *Fragile Dominion: Complexity and the Commons*. Reading, MA: Perseus Books.
Levin, S. A., S. Barrett, S. Aniyar, W. Baumol, C. Bliss, B. Bolin, P. Dasgupta, P. Ehrlich, C. Folke, I.-M. Gren, C. S. Holling, A.-M. Jansson, B.-O. Jansson, K.-G. Mäler, D. Martin, C. Perrings and E. Sheshinsky (1998), 'Resilience in Natural and Socioeconomic Systems', *Environment and Development Economics* **3**(2), 222–235.
Ludwig, D., D. D. Jones and C. S. Holling (1978), 'Qualitative Analysis of Insect Outbreak Systems: The Spruce Budworm and Forest', *Journal of Animal Ecology* **47**, 315–332.
Lutz, E., ed. (1993), *Toward Improved Accounting for the Environment*. Washington, DC: World Bank.
Maddison, A. (2001), *The World Economy: A Millennial Perspective*. Paris: OECD, Development Research Centre.
Majumdar, M. and T. Mitra (1982), 'Intertemporal Allocation With a Non-Convex Technology: The Aggregated Framework', *Journal of Economic Theory* **27**(1), 101–136.
Majumdar, M., T. Mitra and K. Nishimura, eds. (2000), *Optimization and Chaos*. Berlin: Springer-Verlag.
Mäler, K.-G. (1974), *Environmental Economics: A Theoretical Enquiry*. Baltimore, MD: Johns Hopkins University Press.
May, R. M. (1975), *Stability and Complexity in Model Ecosystems*, 2nd ed. Princeton, NJ: Princeton University Press.
May, R. M. (1989), 'How Many Species?', in L. Friday and R. Laskey, eds., *The Fragile Environment*. Cambridge: Cambridge University Press.
Murray, J. D. (1993), *Mathematical Biology*. Berlin: Springer-Verlag.
Pastor, J. and D. J. Mladenoff (1992), 'The Southern Boreal-Northern Hardwood Forest Border', in H. H. Shugart, R. Leemans and G. B. Bonan, eds., *A System Analysis of the Global Boreal Forest*. New York: Cambridge University Press.
Pearce, D. and G. Atkinson (1993), 'Capital Theory and the Measurement of Sustainable Development: An Indicator of Weak Sustainability', *Ecological Economics* **8**(1), 103–108.
Perrings, C., K.-G. Mäler, C. Folke, C. S. Holling and B.-O. Jansson, eds. (1995), *Biodiversity Loss: Economic and Ecological Issues*. Cambridge: Cambridge University Press.
Rahmstorf, S. (1995), 'Bifurcations of the Atlantic Thermohaline Circulation in Response to the Change in the Hydrological Cycle', *Nature* **378**, 145–149.
Rahmstorf, S. (2002), 'Ocean Circulation and Climate During the Past 120,000 Years', *Nature* **419**(6903), 207–214.
Scheffer, M. (1997), *The Ecology of Shallow Lakes*. New York: Chapman Hall.
Scholes, R. J. (2003), 'Convex Relationship in Ecosystems Containing Mixtures of Trees and Grass', *Environmental and Resource Economics* **26**, 559–574.
Scrimshaw, N. C. et al. (1968), *Interactions of Nutrition and Infection*. Geneva: WHO.
Skiba, A. K. (1978), 'Optimal Growth with a Convex-Concave Production Function', *Econometrica* **46**(3), 527–540.

Spurr, G. B. (1988), 'Marginal Malnutrition in Childhood: Implications for Adult Work Capacity and Productivity', in K. J. Collins and D. F. Roberts, eds., *Capacity for Work in the Tropics*. Cambridge, UK: Cambridge University Press.

Spurr, G. B. (1990), 'The Impact of Chronic Undernutrition on Physical Work Capacity and Daily Energy Expenditure', in G. A. Harrison and J. C. Waterlow, eds., *Diet and Disease in Traditional and Developing Countries*. Cambridge, UK: Cambridge University Press.

Starrett, D. A. (1972), 'Fundamental Non-Convexities in the Theory of Externalities', *Journal of Economic Theory* **4**(1), 180–199.

Starrett, D. (1973), 'A Note on Externalities and the Core', *Econometrica* **41**(1), 179–183.

Tirole, J. (1993), *The Theory of Industial Organization*. Cambridge, MA: MIT Press.

van Andel, T. H. (1994), *New Views on an Old Planet: A History of Global Change*. Cambridge: Cambridge University Press.

Wagener, F. O. O. (2003), 'Skiba Points and Heteroclinic Bifurcations, with Applications to the Shallow Lake System', *Journal of Economic Dynamics and Control* **27**(9), 1533–1561.

Weitzman, M. L. (2000), 'The Linearized Hamiltonian as Comprehensive NDP', *Environment and Development Economics* **5**(1), 55–68.

WHO (1983), *Measuring Change in Nutritional Status*. Geneva: World Health Organization.

WHO (1985), *Energy and Protein Requirements*. Geneva: World Health Organization, Technical Report Series 724.

World Bank (1992), *World Development Report*. New York: Oxford University Press.

Scale and Scaling in Ecological and Economic Systems

JÉRÔME CHAVE[1] and SIMON LEVIN[2]

Department of Ecology and Evolutionary Biology, Guyot Hall, Princeton University, Princeton NJ 08544-1003, U.S.A.; [1]*Present address: Laboratoire Evolution et Diversité Biologique, CNRS UMR 5174, 118 route de Narbone F31062 Toulouse, France (E-mail: chave@eno.princeton.edu);* [2]*(E-mail: slevin@eno.princeton.edu)*

Abstract. We review various aspects of the notion of scale applied to natural systems, in particular complex adaptive systems. We argue that scaling issues are not only crucial from the standpoint of basic science, but also in many applied issues, and discuss tools for detecting and dealing with multiple scales, both spatial and temporal. We also suggest that the techniques of statistical mechanics, which have been successful in describing many emergent patterns in physical systems, can also prove useful in the study of complex adaptive systems.

Key words: criticality, ecology, economy, scale, statistical mechanics

1. Introduction

One of the fundamental truths in the study of natural systems is that there is no single correct scale on which to study dynamics. In ecological systems, processes (such as behaviors) at the level of whole organisms represent the collective dynamics of cells (such as neurons) and molecules operating on much faster time scales, largely shaped by evolutionary processes operating over much longer scales of space and time. Similarly, organizations and economies, societies and the Internet exhibit collective dynamics that emerge from the behaviors of individual agents, mediated through levels of interaction and aggregation. The distinctions among scales need not be sharp, as between individuals and their societies, but may represent a continuous gradation, as in atmospheric phenomena, across many orders of magnitude. It is thus of fundamental importance to recognize how our perceptual scales condition the way we describe systems, how patterns change across scales, and how phenomena at different scales influence one another (Levin 1992, 1993).

These pervasive scientific verities become of applied importance when attention turns to the interactions among systems with different dominant scales of activity. The emergence of antibiotic resistance, for example, has become a problem of overwhelming concern because typical bacterial generation times are much shorter than those of humans; and what is perhaps more relevant, resistance has arisen faster than we have been able to produce genuinely novel antibiotics. Other disease-causing organisms, such as *Plasmodium* (the etiological agent of malaria), influenza and HIV have also presented challenges to the development of treatments because of similar mismatches in time scales.

These examples make clear that, although we usually think of evolutionary change in biological systems as operating much more slowly than in socioeconomic ones, the interconnection between biological and socioeconomic systems brings these scales much more into concord. Human activities have sped the evolution of antibiotic resistance, of pesticide resistance in agricultural systems, and of heavy metal tolerance in plants. Harvesting practice, similarly, has influenced the life history dynamics of fish species, selecting for those that mature earlier, and at smaller sizes. Nor do we need to be reminded that human activities increasingly are leading to extinction of species and to changes in the composition of our atmosphere, again accelerating the scales of environmental change well beyond what we have seen in previous centuries of human existence.

Disease management, global change and conservation of Earth's biodiversity are only some of the spectrum of environmental issues that exist at the interface between natural systems and sociopolitical ones. Models can play a fundamental role in informing decision-making, because they allow the integration of processes operating at diverse scales. However, the art of developing models to act across widely different scales of space, time or organizational complexity involves more than just the inclusion of every possible detail, at every possible scale. In the study of general circulation models, for example, building models with finer and finer spatial resolution not only is limited by computational complexity; it also introduces detail that can confound interpretation. The same applies to the analysis of any nonlinear system, in which the challenge in moving across scales is not to include as much detail as possible, but to find ways to suppress irrelevant detail and simplify system description (Ludwig and Walters 1985; Levin 1991; Levin and Pacala 1997). No sensible scientist would try to build a model of the behavior of an individual organism by accounting for the processes within every cell, tied together in a network of interaction of numbing complexity. Similarly, no sensible scientist should try to build models of ecological systems in which one reproduces the behaviors of every organism, or even every species. The goal rather should be to identify relevant detail, and to describe the dynamics of whole systems in terms of the statistical properties of the units that make them up.

Methods of statistical mechanics permit extrapolation from the microscopic to the macroscopic in many physical systems. To describe how water boils or ice melts, or how fluid flows, we do not require detailed information on the atomic structure of the constitutive particles, or their positions and movements. From the rules governing the movement of individual particles, we can derive partial differential equations describing their collective behaviors. Indeed, such models may be mathematically complex, yielding instabilities, chaotic dynamics and nonlinear pattern formation. Nonetheless, they reliably capture the macroscopic behavior in terms of the microscopic displacement of the particles.

In biological and socio-economic systems, deriving such macroscopic equations is also a worthwhile goal, although much more difficult to achieve, despite recent progress (Durlauf 1993; Flierl et al. 1999). First of all, unlike physical systems, the underlying processes are not well-described. Secondly, the range of actions of biological and economic agents, involving rational and not-so-rational decision-making in the face of complex environmental signals, adds a novel form of complexity, beyond the mathematical complexity of relatively simple physical systems. Thirdly, and most vexing, the elements that make up biological and socio-economic systems cannot be well-represented as uniform ensembles of billiard balls. It is the heterogeneity of these systems, and the importance of the continual infusion of new types of unpredictable character, that sets these systems apart. These characteristics define complex adaptive systems (Levin 1999, 2000), in which pattern emerges from the interplay between processes that generate novelty and those that winnow that novelty, based on localized interactions among self-replicating entities of diverse characters. Economies, societies, ecosystems and the biosphere are prototypical complex adaptive systems.

Issues of scaling are fundamentally involved in the analysis of complex adaptive systems, in which macroscopic behaviors of collectives emerge from microscopic interactions among individual agents. In this paper we will direct attention to such complex adaptive systems, across a range of applications. Individuals may interact locally by mechanisms of attraction and repulsion, by exchange of biomass, energy, chemical compounds, financial currencies or information; or by other forms of signaling. These processes engender phenomena at higher levels, such as aggregation into herds or cities, cooperation and warfare, the development of social norms and endogenous organization into societies and religions.

By focusing attention on emergent phenomena, we seek to explicate the mechanisms underlying pattern. It might seem that theories developed for billiard balls might be completely inappropriate as points of departure for describing the collective dynamics of conscious and rational agents, such as humans. Nonetheless, the point is that many phenomena arise as statistical regularities, independent of the fine-scale detail. Collectives then achieve status as super-organisms, whose behaviors may not be conscious in the whole, but may be described as if they were. One often hears market experts, for example, report that "the market was nervous today", or similar expressions. Indeed, the notion that the market has real ceilings and floors is not only an anthropomorphism, but is self-fulfilling if the individual agents that comprise the market believe it. Financial crashes and recoveries, traffic jams, aggregation in animal societies, are typical of the dynamics found in complex adaptive systems.

2. Scaling Laws in Natural Systems

A convenient place to begin in the study of the effects of scale is to ask how one set of variables changes in relationship to changes in others. In the small, of course, that is the stuff of the differential calculus; but scaling laws seek to capture those relationships across a range, and in terms of compact analytical relationships. Brock (1999) defines a scaling law as a "common property of a set of plots of one quantity against another". He furthermore reminds us that Gell-Mann (1994, p. 97) attributes the admission to Mandelbrot that "quite frankly ... early in his career he was successful in part because he placed more emphasis on finding and describing the power laws than on trying to explain them".

In many applications, the search for scaling relationships has focused on length or time scales as the primary variables, and asked how other variables change in relation. Mandelbrot (1975) has illustrated the power of such analyses, demonstrating for example how the measurement of the perimeter of a land mass, or a snowflake, or a cloud may vary in relationship to the length of the measuring stick used. Following Mandelbrot (1975), we may ask what the length L_ε of a coastline, when measured with a measuring stick of length ε, is. This length generally increases steadily as ε decreases, suggesting that L_ε only makes sense if one has specified the scale first. The relation between L_ε and ε is of the form $L_\varepsilon \approx \varepsilon^{-d_f}$ as ε tends to zero, which allows us to define the fractal dimension $d_f = \lim_{\varepsilon \to 0} \frac{\ln(L_\varepsilon)}{\ln(1/\varepsilon)}$ of the set (Hastings and Sugihara 1993). It is possible, and indeed quite easy, to give rules for the construction of "self-similar" sets, for which the above scaling law $L_\varepsilon \approx \varepsilon^{-d_f}$ is in fact valid at all scales. An example is von Koch's snowflake (Figure 1, Mandelbrot 1975). The existence of such relationships implies that geometric concepts such as perimeter may not be absolutes, but make sense only relative to the length scale used, so that there are no natural scales that must be employed. Such scale-free phenomena are ubiquitous in Nature, and beg explanation. They are characteristic of critical phenomena in physical systems, but can also arise in other ways, as for example in the development of bronchial networks (Barabási et al. 1996; West et al. 1997).

In ecological systems, such scaling relationships in regard to length (or area or volume) are among the most robust empirical generalizations found. Particularly influential have been allometric relationships, which govern the dependence of variables such as metabolic rate on body size; species-area curves, which relate the number of species found to the area of study; and self-thinning laws, which for example relate the total biomass of a forest to the mean size of the trees.

The relationship between metabolic rate M and body mass B is a well-documented one, often fitted by a power-law function (Calder 1984; Peters 1983, see Figure 2):

$$M \approx aB^b$$

When plotted on log-log paper, M is a linear function of B, and the straight line has a slope b. The coefficient a varies across the life forms, and the exponent b between

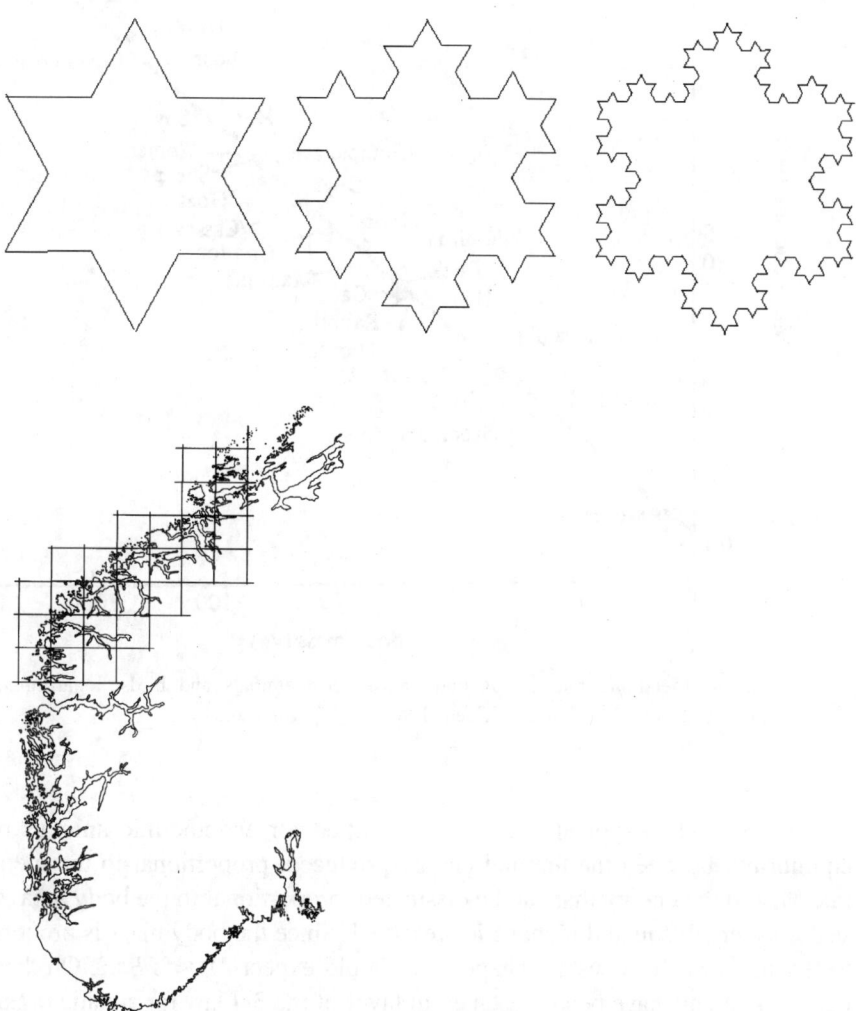

Figure 1. Two classic examples of fractals. *Top*: first three stages in the construction of von Koch's snowflake fractal (from left to right). The total area increases as the procedure iterates. Assuming that the first-stage figure has an area A0, the second-stage triangle has an area $A_0 + 3A_1 = A_0 + 3/9 A_0 = 4/3 A_0$. More generally the nth-stage figure has an area $(1 + 3/5(1 - (4/9)^{n+1}))A_0$, which tends to $8/5 A_0$ as n goes to infinity. The perimeter, however, is unbounded, as its length is multiplied by 4/3 from one stage to the next. The fractal dimension of the von Koch fractal is $\ln(4)/\ln(3)$. *Bottom*: fractal structure of the coastline of Norway.

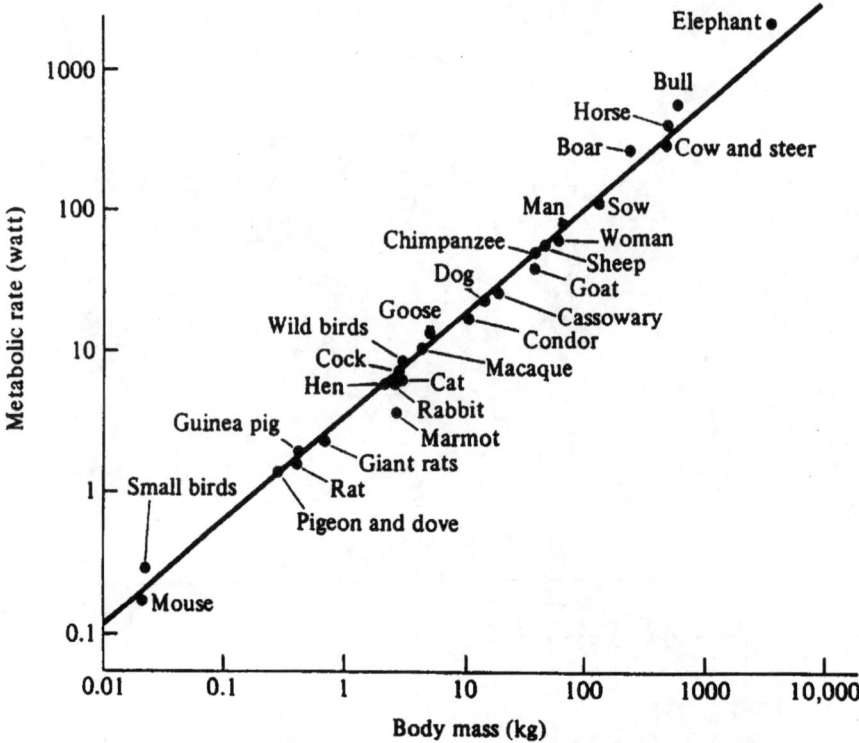

Figure 2. Metabolic rate versus body mass for mammals and birds, logarithmic axes. Reproduced from N. Schmidt-Nielsen (1984), with permission.

2/3 and 3/4. The exponent 2/3 can be justified for endothermic animals by the equilibrium between the thermal energy produced, proportional to the metabolic rate M, and the energy that can be dissipated, proportional to the body area, which scales as length squared if shape is preserved. Since the body mass is proportional to length cubed for constant shape, one should expect $M \approx aB^{2/3}$. On the other hand, arguments have been presented in favor of the 3/4 law for aquatic organisms (Niklas 1992), and more generally (West et al. 1997). We shall return to these arguments when we discuss scaling laws from an evolutionary perspective.

Trees have been much studied in the context of allometric relationships. The relation between tree height and stem diameter, for example, is mechanically constrained. Indeed, a tree cannot only grow in height (primary growth), for the stem would be unable to prevent it from buckling. The maximal height H reached for a stem diameter D can be found by means of dimensional analysis. The torque exerted on the tree by wind, or some other destabilizing factor, is proportional to H^3, while the elastic force is proportional to the cross-sectional area of the stem, D^2. Equating these two forces yields $H \approx aD^{2/3}$. In competitive environments, such as tropical rain forests, trees often attempt to maximize their height and their

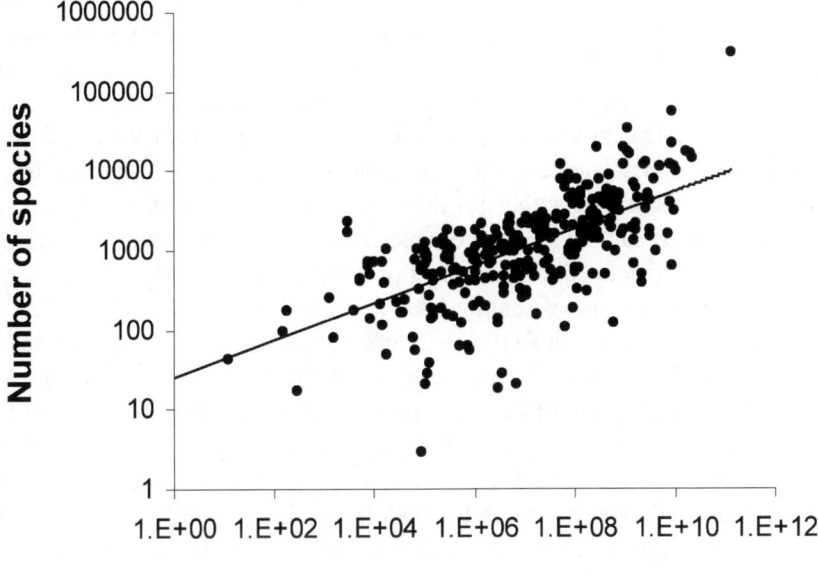

Figure 3. Species area curve for vascular plant species, for 316 islands, regions, or continents (after Williams 1964; Major 1988; Prance 1978, 1994). The rightmost point is a recent estimate of the number of vascular plants on Earth (320,000 species). Note that both axes are logarithmic. A best fit with a power-law function gives $S = cA^z$, with $z = 0.23$. The apparent convexity of the dataset is probably due to the under-representation of the tropical biome.

form factor is only limited by buckling (O'Brien et al. 1995). A similar allometric relationship has also been found for the largest trees in the U.S. forests (McMahon 1975). Likewise, a relation between tree mass and diameter of the form $B \approx aD^b$ has been suggested. While a simple dimensional analysis suggests that b should be 3, Enquist et al. (1998) have shown that the exponent 8/3 may be more appropriate. Empirical analysis using more than 1,800 tropical trees yields a value of $b = 2.55$ (Chave et al. 2001).

Species-area curves probably are the most famous empirical generalization in ecology. It has been long observed that the number of sampled species increased when the survey area was increased. In the beginning of the twentieth century, Arrhenius (1921) suggested that the allometric dependency of species number with area should follow a power law $S \approx cA^z$. Since then, an impressive body of studies has sought to test this relationship in a large variety of systems and have found values of z ranging between 0.1 and 0.4 (Connor and McCoy 1979; Rosenzweig 1995, Figure 3). Such power-law relations suggest self-similarity, and provide attractive generalizations (see for example, Harte et al. 1999). Yet, recent studies (Condit et al. 2000; Plotkin et al. 2000; May and Stumpf 2000; Crawley and Harral 2001) question the universality of these relations.

In physics, scaling laws have similarly received much attention. Ohm's law, which predicts a linear scaling of the electrical potential when the electrical intensity through a conductor is varied, has never been rigorously proved, but emerged as a robust experimental result during the nineteenth century. The critical behavior of physical systems near a transition point has also yielded important experimental results. The susceptibility, a measure of how sensitive the system is to small perturbations near the transition, scales with the distance to the transition. Examples of such systems are water near the boiling point, iron becoming spontaneously magnetized, or the transition to superconductivity.

In economics, the use of scaling laws has also been highly informative, especially since the work of Zipf (1949). Zipf's law states that city size decreases in inverse proportion to its rank (Makse et al. 1995). Other power laws have been applied to the distribution of the sizes of firms (Gibrat's law, see Ijiri and Simon 1977; Amaral et al. 1997; Brock 1999), of personal income (Pareto's law) and of stock market gyrations. Indeed, the scaling of market fluctuations with the size of the temporal window is the basis for options pricing. Pareto's law (Pareto 1896; see Mandelbrot 1963) states that the density of individuals with incomes greater than u satisfies the form $f(u) \approx cu^{-\alpha}$, where α is a scaling exponent. MacAuley (1922) criticized Pareto's generalizations on several grounds, pointing out that the exponent varies among countries, and disputed its validity for small incomes. Still, it remains a popular notion.

3. Detecting Scales

Finding the relevant scale of description is an important step in the theoretical analysis of a natural system. Usually, only a few scales are relevant, each corresponding with one process that drives the dynamics of the system. In many cases, the investigator must perform rather sophisticated analysis of data to detect the relevant scales in the system of study, and we give a few examples of these techniques.

Detecting relevant scales is of crucial importance in areas such as econometrics. If one is able to find the regularities in financial data, an advantage will be gained in attempting to predict the future evolution of the market. The science of financial markets has witnessed a very rapid development during the 1990s (for an introduction, see Mandelbrot 1997; Mills 1999; Bouchaud and Potters 2000). Financial markets are a good example of complex adaptive systems, and data are widely available. Figure 4 (top, left) displays Standard & Poor's 500 financial index (daily data normalized by the annual average). We use these data to present some of the techniques used to detect relevant temporal scales. Clearly, this signal is not purely random, as one can see from the histogram (Figure 4, bottom, right), which is skewed to the right.

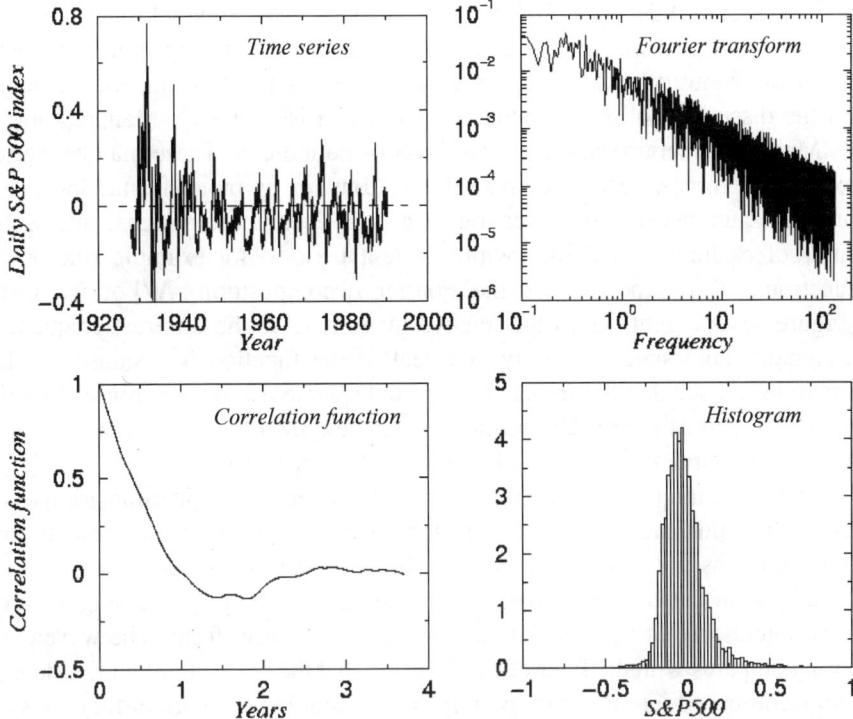

Figure 4. Detecting the scales in the Standard & Poor's 500 financial daily index, from 1928 to 1997 (data from Mills 1999). Left, top panel: standardized normalized S&P500, $(x - \bar{x})/\bar{x}$, where x is the daily index and $\bar{x}$ is the annual index (running average). Right, top panel: Fourier spectrum $S(\omega)$ of the time series. The spectrum decreases with frequency (i.e., large fluctuations are more rare) like a power law $S(\omega) \approx \omega^{-0.93}$. Left, bottom panel: correlation function for the temporal signal. The correlation has a characteristic time of about a year. Right, bottom panel: histogram of the temporal signal. The histogram is obviously skewed towards the right, and non-normal.

In order to detect temporal scales in this signal, one usually uses the correlation function:

$$C(\tau) = \langle f(t) f(t + \tau) \rangle - \langle f(t) \rangle^2,$$

where the brackets denote the time average. $C(\tau)$ gives the probability that the signal is identical at two instants separated by a time lag τ. The function does not depend upon t if the studied time process is stationary. In Figure 4 (bottom, left), we have plotted $C(\tau)$ normalized by the variance of f, $\langle f(t)^2 \rangle$, so that $C(0) = 1$. Many processes have a "finite memory"; that is, the corresponding correlation function falls off exponentially as τ goes to infinity. For these types of processes, one defines the correlation time as

$$C(\tau) \approx \exp(-\tau/T)$$

For the S&P500 data, we find $T = 421$ business days. Beyond this time, one can assume that the events are perfectly uncorrelated. For time lags much smaller than T, on the contrary, one can assume that there has been no loss of memory. For further discussion of scales in finance, the reader is referred to Mandelbrot (1997).

Many time series show a certain level of periodicity. These may be caused by exogenous forces, such as climate fluctuations, or by biotic fluctuations, such as prey-predator cycles. An appealing idea is to decompose a signal into weighted sums of oscillating functions with different periods, for example sine or cosine functions. This procedure is the Fourier decomposition. A Fourier spectrum (Figure 4, top, right) gives the relative magnitude of the different frequencies in a signal. There should be only one peak if the function is a sinusoidal. In our example, we see that the frequencies have the same magnitude from 0.1 to around 0.4 (in year units, 1 yr = 270 business days), and then fall off as a power law. This suggests a first temporal scale of around 100 business days.

These techniques also apply for spatial patterns. Fourier transforms can be used for multidimensional data, or along a predefined direction. Spatial correlation functions are also easily defined. In fact many other techniques have been developed in the context of pattern recognition and image processing. One of the most interesting of these techniques is the wavelet transform. The wavelet transform compares with the Fourier transform in that the data (often multidimensional) are decomposed into a linear combination of simple orthonormal functions $\phi_i(x)$; that is, functions such that $\int \phi_i(x)\phi_j(x)dx = \delta(i - j)$, where $\delta(i) = 0$ if $i \neq 0$ and $\delta(0) = 1$. However, instead of using sine or cosine functions to form an orthonormal basis as in the Fourier transform, wavelet transforms use functions parameterized by a scale factor. In other words, while the Fourier spectrum informs one about the magnitude of a certain frequency in the pattern, the wavelet spectrum informs about the magnitude of a certain scale in this pattern. Therefore, this is a very natural tool to detect spatial scales in an image.

4. The Adaptive Significance of Scale

We now turn our attention to the evolutionary consequences of scale in biological systems. Mechanical and thermal constraints for maintenance of life and for reproduction of biological organisms are at the core of the theory of natural selection. D'Arcy Thompson first brought forward the relation between design and mechanical constraints for organisms. The shape of a bone and the size of a tree are the result of evolutionary optimization within a game-theoretic framework. Observed allometric relationships therefore are the products of evolution as well, and expose implicit tradeoffs and constraints.

Following this idea, authors such as Niklas (1992) and West et al. (1997) take an engineering approach to explaining observed allometric relationships, with remarkable success. They hypothesize appropriate constraints and optimization principles (in terms of efficiency), and from them derive observed relationships. For trees, one constraint is that the tree should be area-preserving; that is when it branches, the cross-sectional area before the branch equals the total cross-sectional area beyond the branch (Shinozaki's pipe model, Shinozaki et al. 1964; Enquist et al. 1999). However, such a choice of constraints can be criticized (Horn 2000), and comparison with real data is often difficult (Schmidt-Nielsen 1984). In addition, as Jacob (1977) has pointed out, evolution does not optimize, but tinkers; furthermore, all evolution takes place within the context of other types, which argues for game-theoretic approaches. Despite these caveats, this theory has been astoundingly successful.

4.1. SCALE AND THE EVOLUTION OF DISPERSAL

Environments vary and the degree of variation depends upon the temporal and spatial window of observation. A wise investor considers this in her strategies, and manages portfolios by widening the window of investments to spread risks. Evolution takes a similar path to reduce the risks that organisms face, for example by spreading risks through dormancy or dispersal.

The reproduction of plant species is highly limited by the dispersal stage. Plant species exhibit a vast spectrum of dispersal strategies to address this problem. Some species invest in a few large seeds, dispersed nearby the parent plant, while some species produce a tremendous amount of seeds that can be dispersed far. In a tropical forest of French Guiana, the smallest seeds are less than a milligram (e.g., *Cecropia* spp., Cecropiaceae), while some are well over a kilogram (*Couroupita guianensis*, Lecythidaceae). The diversity of strategies is evidence of the fact that there is no single optimum strategy, but rather that adaptations occur within a game-theoretic framework.

While most propagules are dispersed near the parent plant, a few centimeters for weeds to a few meters for trees, some are dispersed far away. Short-distance dispersal is mostly driven by gravity. In contrast, long distance dispersal is mostly mediated by animals (Fragoso 1997), wind (Nathan et al. 2000) or water (Whittaker et al. 1989). At the plant scale, the adaptation of the dispersal mode to environmental constraints is crucial for reproduction. At the stand scale, dispersal patterns control the spatial distribution of species and their coexistence (Hubbell 1997; Pitman et al. 1999; Chave et al. 2002). Short range dispersal is crucial to population persistence at this scale.

At the landscape scale, dispersal controls the dynamics of ecological succession after major disturbances. For example, after the last glacial stage (18,000 radiocarbon years before present), forest tree species migrated northwards in North America. The rate at which this process occurred has puzzled scientists for decades, because they assumed that the expansion of the range of a species should occur gradually, as a result of multiple short-distance dispersal events (Skellam 1951). The rapid spread, however, has its explanation in rare occurrences of long-distance dispersal events as evidenced by Clark and collaborators (1998). In fact, the importance of long distance dispersal had been perceived much earlier for the founding of new genetically isolated populations (Wright 1943), for the spread of epidemics (Mollison 1995), and for the dispersal of pest species (Andow et al. 1990).

At the regional and continental scales, seed dispersal is an important factor in determining plant species distributions. For example, many neotropical plant species are endemic to a sub-region. Gentry (1992) estimated that, in each of the 9 phytogeographic regions in Central and South America, plant endemism was greater than 50% (except in the coastal zone of Colombia). This endemism cannot be explained only by the variety of environmental types, as evidenced by Prance in several speciose families such as Chrysobalanaceae (Prance 1973). In principle, mathematical models of long distance dispersal allow for dispersal events over any scale. However, the frequency of large-scale dispersal events is very low. If a seed is dispersed hundreds of kilometers away from any conspecific tree, and it does not die in the juvenile stage, this individual will be the founder of a new population, genetically isolated from the parent population, and therefore likely to initiate speciation (Brown and Lomolino 1998; Nathan 2001).

Dispersal is a form of exploration conditioned by the scales of variation in a system. There is direct analogy here with the investment strategies or the degree of risk-taking in the management of organizations. Innovation is advantageous in a changing environment; how advantageous depends upon the scales of environmental variation.

4.2. SCALE AND MARINE ORGANISMS

Life in (moving) fluids presents fascinating challenges (Vogel 1994). Swimming organisms span a huge range of body sizes, from bacteria to whales. We shall focus on the evolutionary significance of this variation in the next section; but for the moment, we seek to understand what it means for the organism, to live in a moving environment.

The dynamics of fluids are fully described by a well-known set of partial differential equations, the Navier-Stokes equations (Batchelor 1953). One can accurately predict the flux of water transported though a pipe when the water velocity V is known. Yet, as the velocity increases, the flow becomes turbulent. The reason for this transition is that two processes are operating at the same time. At small velocity, the momentum flux carries water in the pipe along linear streamlines (laminar flow). At high velocities, however, the laminar flow is destabilized by frictional forces. Dimensional analysis of the Navier-Stokes equations, which govern the dynamics of the fluid, gives an insight to this problem. To illustrate this, consider the Navier-Stokes equation, in one dimension, in the absence of gravitational effect or pressure:

$$\frac{\partial V}{\partial t} + V\frac{\partial V}{\partial x} - \nu\frac{\partial^2 V}{\partial x^2} = 0$$

in which t is time and ν is the viscosity (coefficient of friction). This equation describes the rate of change of velocity, $\partial V/\partial t$, as a function of the spatial profile of this velocity. The momentum flux is the term $V\partial V/\partial x$, and has the dimensions of V^2/L, where L is the cross-section of the pipe. If x is measured in meters and V in meters per second, the frictional term is $\nu \partial^2 V/\partial x^2$, which has the dimensions of $\nu V/L^2$ (ν has units square meters per second). The ratio of these two quantities is a basic scaling parameter, the Reynolds number

$$\text{Re} = \frac{V^2/L}{\nu V/L^2} = \frac{VL}{\nu}$$

Re has no dimension. A laminar flow is observed for Re < 2,000 in a typical water pipe, while turbulence is usually observed for Re > 40,000.

The Navier-Stokes equations are simply balance equations. One can incorporate many other processes that drive the fluid dynamics, such as the change in pressure (this would be an additional term $1/\rho \partial P/\partial x$, where P is the pressure and ρ the density), or the gravity ($-g$, which is the acceleration constant). Therefore, the relative magnitude of the other processes at work can be summarized into other non-dimensional numbers, like the Reynolds number.

The analysis of the Navier-Stokes equations in the turbulent regime poses a formidable mathematical challenge. An intriguing property of the turbulent regime is that eddies dissipate the energy over a broad spectrum of scales. This is often described in a temporal context, since air sensors are static. The energy-frequency spectrum $E(\omega)$ – that is, the energy dissipated for a given frequency ω, – is (Kolmogorov 1941; Batchelor 1953; Frisch 1995)

$$E(\omega) \approx \omega^{-5/3}$$

This scaling relationship can be translated in a spatial context, since large eddies are those that are rare (low frequency ω) and that therefore dissipate the largest amount of energy $E(\omega)$. The Kolmogorov – 5/3 scaling law has been observed in a large number of real fluids. The turbulent regime is chaotic: in a turbulent fluid, two configurations infinitesimally close initially diverge exponentially fast. Although the system is governed by deterministic (Navier-Stokes) equations, the rapid divergence of initially similar states occurs. To further illustrate the difficulty of analyzing the Navier-Stokes equations, let us mention that the existence or the non-existence of "blow-up" solutions in the three-dimensional problem (tornado-like solutions) is still not established, and recent work suggests that no such singular solution exists (Ya. Sinai, unpublished results). The study of turbulent fluids is of paramount importance for the current challenges of oceanic and atmospheric sciences. Indeed, these equations govern the dynamics of both environments and they constitute the core of most globally coupled models of ocean-atmosphere interaction (Stull 1988).

Obviously, it makes a big difference whether the surrounding environment is turbulent or if it is not for a swimming organism. Since both velocity and size increase with body weight, minute organisms live in a laminar and indeed quite viscous liquid, while fish live in a turbulent fluid. Sub-millimetric animals displace themselves using flagella; they use the same technique as snakes on the ground. This is not to say that these animal cannot move large distances; they can easily be carried by streams. In contrast, large fish have a highly optimized design. A most important optimization constraint, in this case, is the minimization of the energy needed to displace the fish's body, a quantity called the drag. This is why fish have an aerodynamic shape, and well-designed body surfaces (Vogel 1994). Constraints are different for flying animals. Drag is not a crucial issue, because of the low density of air, but they must overcome gravity. Flying devices are wonderful instances of evolutionary design. The technical solutions are varied, from insects propelled by small wings, flapped at high frequency, to large birds, which take full advantage of wind upwelling streams.

So far, we have considered the consequences of living in a moving fluid for a single organism; but at the level of a community of organisms, environmental constraints also have deep implications. For example, birds often assemble in geometric, vee-shaped groups, during their migration flight. This technique reflects community-level optimization: birds reduce their energy expense globally by this formation. This emergent pattern is therefore intimately related to the physical properties of the environment. Numerous examples reinforce this observation for other organisms, such as the clustered large-scale distribution of zooplankton in the ocean, likely due to a response to physical or chemical conditions (Levin et al. 1989), or the multi-scale distribution of tree species in tropical rainforests (Plotkin et al. 2002). Importantly, the explanation for observed spatial distributions may differ across scales, say with exogenous forces dominating on broad scales, and endogenous forces on small scales. Spectral analysis shows a $-5/3$ scaling on broad scales, typical of turbulence, but much flatter distributions on finer scales (Levin et al. 1989).

Spatial aggregation is a property of interest not only to ecologists, but to economists and other social scientists as well. Recently, Flierl et al. (1999) used numerical techniques coupled with methods of statistical mechanics (see section 5) to investigate the role of local interaction among organisms as an important cause of pattern formation in communities of marine organisms. We return to this in section 6.

5. Power Laws in Physical Systems: How Do They Arise?

A number of theories have attempted to provide a better understanding of the conspicuousness of power laws in Nature. Most of them, for example the idea of self-organized criticality, are closely related to the fundamental concept of critical transitions.

5.1. CRITICAL SYSTEMS AND RENORMALIZATION

When one heats water to the boiling point, or cools steel down to the Curie temperature, the system undergoes a dramatic transition, whence the term "critical". The state of water changes from liquid to gas, whereas steel below its Curie temperature has a spontaneous magnetization (Stanley 1971; Yeomans 1992; Binney et al. 1992). In this case, the transition is from an ordered state (liquid, ferromagnet) to a disordered state (gas, paramagnet). In the words of Kauffman (1993), the transition point is a "sweet spot" between a too-ordered system and a too-disordered one. It has been observed experimentally that, in the vicinity of the transition, many physical quantities vary according to a power-law as the parameter varies (Stanley 1971; Binney et al. 1992). A conceptual breakthrough in the study of these systems was the realization that although a vast number of such transitions are observed in Nature, they all fall into a few classes, referred to as universality classes.

The simplest example of critical transition is encountered in porous media or fractured rocks (Broadbent and Hammersley 1957). The question is "What is the chance that water will percolate through the medium?" Obviously, this question is related to the fraction of accessible pores in the medium. To model this system, it is helpful to think of a lattice with open and closed bonds. Let us consider a taxi driver in Manhattan, New York. He wants to drive from the upper side to the lower side, without respect for the one-way signs, and without knowing its direction beforehand. However, $1 - p$ of the streets are jammed. Then, it can be shown that the taxi driver succeeds in his enterprise if only if $p \geq 1/2$ (Kesten 1980). This result actually holds only in the limit of an infinite system size. The probability of percolation, when p is just slightly larger than $1/2$, behaves as $(p - 1/2)^\beta$, where $\beta = 5/36$, and a set of exponents characterizes this transition. At criticality, that is at $p = 1/2$, the set of occupied bonds is fractal, and the fractal dimension is $d_f = 91/48$. A very useful approximate scheme makes full use of this scale invariance: It is the renormalization procedure. Since the system is scale-invariant, its properties at criticality should not be much changed when the scale unit is varied. Therefore, let us take a scale twice as large (Reynolds et al. 1977). In Figure 5, we show how open and closed bonds can be redefined by using a larger unit of measurement. After the transform, blocks of 2×2 bonds are considered as a single bond, either open (if the underlying 2×2 network is open) or closed. Thus, knowing the fraction of open bonds p_1 at scale $k = 1$, we find the fraction of open bonds p_2 at scale $k = 2$. The relationship is in general valid from scale k to scale $k + 1$, and it reads

$$p_{k+1} = p_k^5 + 5p_k^4(1 - p_k) + 8p_k^3(1 - p_k)^2 + 2p_k^2(1 - p_k)^3$$

The equation here deals explicitly with scale, for it relates the state of the system at scale k to its state at scale $k + 1$. The key observation, made by Kadanoff (1966), is that the system at criticality should be scale invariant, so that the critical p should be a fixed point (i.e., a stable root) of the above equation. The only fixed point of this equation is at $p = 1/2$. This threshold usually depends on the choice of

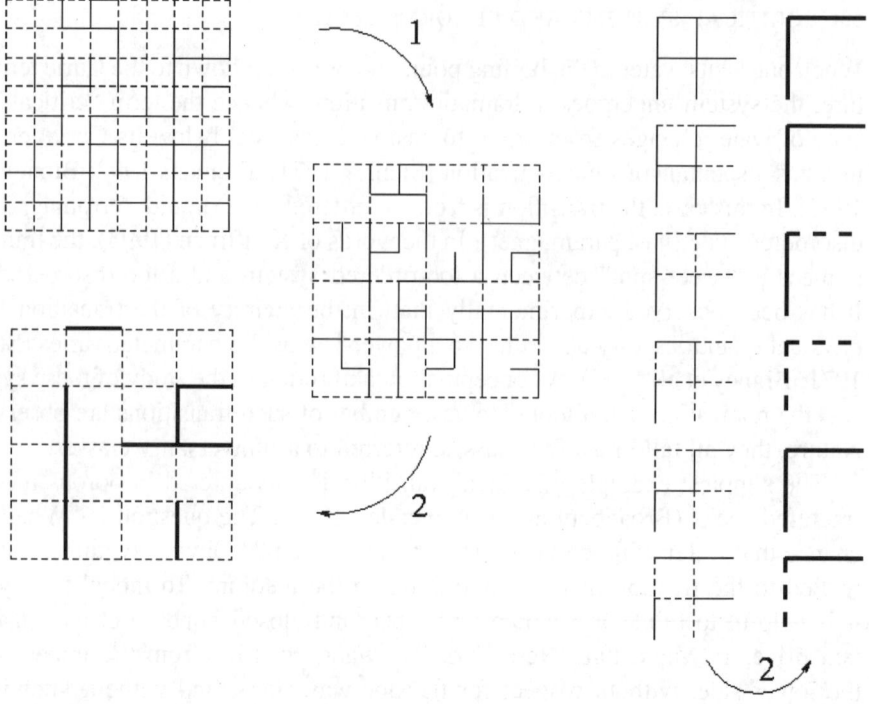

Figure 5. Description of the real-space renormalization procedure for the bond percolation problem used by Reynolds et al. (1977). Left: step 1 corresponds to a change in scale, from scale 2^k to scale 2^{k+1}. Step 2 corresponds to the renormalization procedure. During this procedure, an eight-bond cell is replaced by a two-bond cell. Right: examples of the renormalization of a few of the possible 8-bond cells. The renormalization rule is identical for the horizontal bond and for the vertical bond: if there exists an open path from the left to the right then the renormalized horizontal bond is open.

the renormalization scheme; the fact that p coincides with the exact threshold is fortuitous. One can also use this equation to estimate a wealth of critical exponents. Moreover, a fascinating correspondence between the percolation problem and a model of dynamic critical transition, the Ising model, has been found (Fortuin and Kastelyn 1972). The study of the static model of percolation therefore provides valuable information on the equilibrium state of a dynamical system.

The techniques of the renormalization group were first developed as a technical tool in quantum field theory (Bogoliubov and Shirkov 1959), where most of the theory can be dealt with in an abstract space (Fourier space). In contrast, Kadanoff's renormalization is referred to as "real-space" renormalization (Wilson and Kogut 1974). The notion of critical transitions is of great potential interest, both in ecology and in economics. In both disciplines, the spatial (or more abstract) organization of individual agents into interacting networks can determine the emergence of patterns (cities, aggregations of animals, human societies, trophic webs) and thereby influence fundamentally the flows (capital, information, energy, disease). Thus, extensions of the Ising model, such as interacting particle systems, have become the objects of quantitative study in examining these problems (Durrett and Levin 1994).

5.2. SELF-ORGANIZED CRITICALITY

As just discussed, empirical studies have shown that when a parameter (temperature, occupation of the percolation lattice, reproduction rate) is tuned near the transition point, these systems become insensitive to the scale of study; that is they become scale-invariant. Bak, Tang and Wiesenfeld (1988), have suggested that complex systems of interacting agents "evolve naturally towards a *critical* state, with no intrinsic time or length scale". For example, a pile of sand constructed by slowly adding one grain of sand at a time, eventually reaches an angle for which the addition of one grain can cause an avalanche. They have sought to apply this idea to systems as diverse as earthquakes (Sornette and Sornette 1989), the species extinctions observed in the fossil record (Bak and Sneppen 1993), and forest fires (Malamud et al. 1998). Kauffman (1993) has suggested that such concepts should also apply to economic and social systems. Ecosystems and societies, however, are not sandpiles; they are made up of heterogeneous assemblages of individual agents interconnected through irregular networks. Although valuable as a concept, self-organized criticality is too simplistic a notion for most natural systems (Levin 1999).

5.3. CRITICALITY IN ECOSYSTEMS

The classical model of the dynamics of a single species is the Verhulst equation

$$\frac{du(t)}{dt} = bu(t) - mu(t) - au(t)^2$$

In this, individuals produce offspring at a rate b, and die at a rate m ($b > m$). The population reaches its carrying capacity $u^* = (b - m)/a$, after an initially exponential growth. Such a model also applies to the growth of the population of a city. The borderline region where the reproductive ratio $R_0 = b/m$ is close to one is often observed in real communities. Exactly at $R_0 = 1$, the population undergoes a critical transition and exhibits non-trivial properties. For example, $P(t)$, the population's survival probability, decreases as $P(t) \approx 1/t$. In addition, for the model on a lattice with limited distance dispersal, R_0 should be at least $R_0^c > 1$ (Harris 1974), and it is observed numerically that

$$P(t) \approx t^{-\delta}$$

where $\delta = 0.451$ in two dimensions of space. Note that the non-spatial theory predicts $\delta = 1$. Also, starting from a small population cluster, the number of individuals grows over time as $N(t) \approx t^\theta$, with $\theta = 0.230$. One may also wonder how an initially small population cluster spreads over time. The radius $R(t)$ (averaged over surviving populations) also grows as a power law of the time $R(t) \approx t^{1/z}$, with $z = 1.73$ in two dimensions ($z = 2$ in the non-spatial model). The three exponents δ, θ, and z are independent of each other, and are related to other exponents, such as the fractal dimension of the set of occupied sites, $d_f = z\theta/2$. A recent review of this theory is given in Hinrichsen (2000). Note that the simple on-lattice population model depicted above also undergoes a critical transition around $R_0 = R_0^c$, belonging to the so-called "directed percolation universality class". This universality class is characterized by a set of scaling exponents, among which are δ, θ, and z. Indeed, virtually any quantities satisfy power laws near the critical point, and the behavior of many quantities in the vicinity of the critical point are given by functions of $(R_0 - R_0^c)$. These exponents take non-trivial values in low-dimensional system (dimension $d = 1$ to $d = 4$), and they are related by a relationship between exponents, referred to as the hyperscaling relation (Grassberger and de la Torre 1979)

$$\theta = \frac{d}{z} - 2\delta$$

When more than a single species is present in the community, interactions drive the important processes, and shape community-scale patterns. In a real community, competition interacts with dispersal and differential mortality patterns in a nontrivial way. To test for all possible combinations of processes, factorial experiments can in theory be set up, but these are difficult. Models provide an alternative insight into this problem. Early models of species coexistence were built upon Lotka-Volterra equations, which describe the competition for resources. One of the main conclusions of this approach is that two species feeding upon the same resource cannot coexist. A number of theories have been offered to explain this apparent problem. We have already seen that at the largest scale, dispersal and speciation play an important role, together with environmental variations (climate, rainfall, matrix type). At such large scales, interspecies competition is irrelevant. It becomes important, however, at smaller scales. A simple individual-based model such as the one discussed earlier can serve as a basis to address the issue of species richness and to provide theoretical predictions for the species-area curves. In the simplest model, it can be shown that the species with the largest reproductive ratio always drives all other species to extinction. However, if a tradeoff is incorporated, the species with largest reproductive ratio also is the poorest competitor. In this case, a non-trivial equilibrium is reached (Levins 1969; Buttel et al. 2002; Chave et al. 2002). Figure 6 shows the spatial arrangement of coexisting species at different instants of the simulation. A power-law species-area relationship is observed. We observe that this relationship does not come from the critical nature of the system. Other scaling laws for abundance distributions have also been derived (e.g., Kinzig et al. 1999).

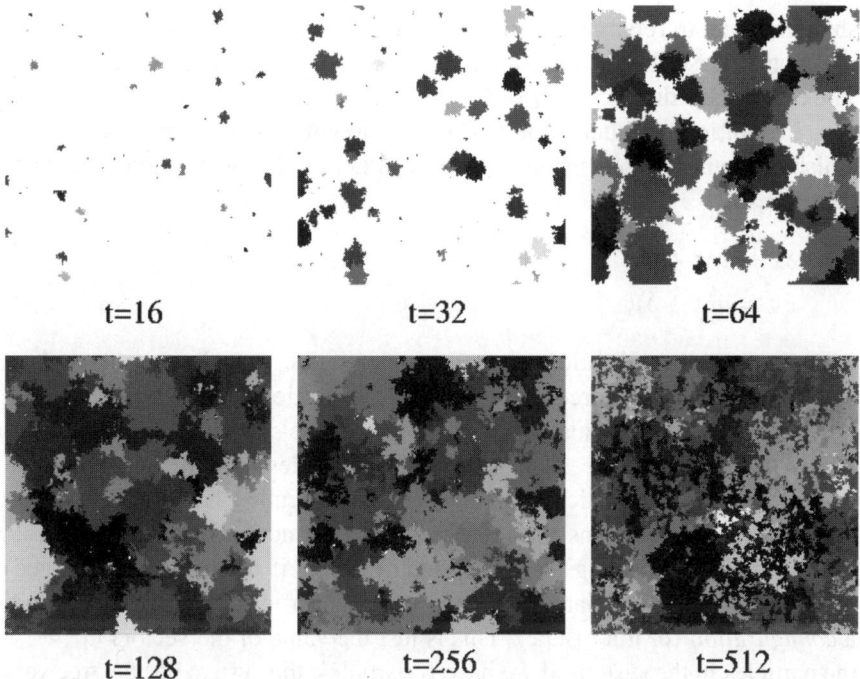

Figure 6. Output of a spatially-explicit model of biodiversity: the tradeoff model. Each species possesses a different death rate and the same birth rate. Species with the smallest death rate are at an advantage over the other. However, these species are poorer competitors: when they compete for a site with a species with a larger death rate, they always loose. At early times (t = 16 to t = 64), there is little interspecific competition, and monospecific domains grow as in the supercritical contact process. However, at later times, competition becomes important, and strategies with a large death rate, but with little habitat limitation, are increasingly often observed.

6. Statistical Mechanics and Kinetic Theory

The utility of the concepts developed in the previous section make one may wonder whether these could apply to other systems, which lack such a well-defined mathematical representation. A crucial remark is that the large-scale patterns observed in complex adaptive systems such as markets, social interactions, and ecosystems are generally not the mere superposition of an assembly of organisms, but the result of interactions among these "agents". This situation is reminiscent of what happens in a gas: each particle performs a complex displacement, changing its velocity and orientation during the interactions with other particles. However, the gas as a whole has well-understood properties, and these properties are not directly related to the history of each particle in the gas. For example, the temperature of a gas is a purely macroscopic concept, and it does not apply at the scale of a particle.

We here briefly sketch the method of statistical mechanics used to find the macroscopic kinetic equations of an assembly of interacting organisms, and we shall continue to refer to the example of gas particles. We assume that the particle is punctual, described by a position vector x and an instantaneous velocity vector v. The elementary displacement of the particle, δx, depends on the value of the velocity. In addition, an interaction of the particle with its environment is recorded by a displacement $(\delta x', \delta v')$ from the position (x, v). The total microscopic equation therefore is:

$$\delta x = v\delta t + \delta x_e \qquad (1)$$
$$\delta v = a\delta t + \delta v_e$$

where a is the acceleration of the particle. If the particle is solitary, Newton's equation $F = ma$ (force exerted upon the particle equals the particle's mass times its acceleration) closes the description of the system. The external displacements $(\delta x_e, \delta v_e)$ can be taken as random variables, and we are left with a set of two coupled stochastic differential equations, called Langevin equations. The vector $(\delta x_e, \delta v_e)$ contains the information of the action of other particles, while F describes other forcings (e.g., an electric force, if the particle is charged). The global description of a large assembly of such particles requires the definition of a *configuration* (or microstate). This is just the value of the vectors (x_i, v_i), for all the particles in the system. If we have N particles, the system has $6N$ free variables, and a configuration is a point in a $6N$-dimensional space called the phase space.

The heart of the kinetic theory consists in writing down the balance equation for the number of particles contained in a small volume of this phase space (cf. Appendix). The resulting differential equation can then transformed into Boltzmann's equation by neglecting collision terms $(\delta x_e, \delta v_e)$ in equations (1).

In a more general setting, forms of the Fokker-Planck equation can be deduced directly when the particles are considered as random walkers, i.e., when $\delta V_e = 0$. A large assembly of – interacting or non-interacting – random walkers possesses important macroscopic properties. In the non-interacting case, the dynamic equation (equation 4, appendix) becomes:

$$\frac{\partial n(X,t)}{\partial t} = \frac{1}{2} \sum_{\alpha,\beta} \frac{\partial}{\partial X_\alpha \partial X_\beta} D_{\alpha\beta} n(X,t) \qquad (5)$$

which simply is the diffusion equation. The matrix $D_{\alpha\beta}$ is the covariance matrix $\langle \delta X_{e\alpha} \delta X_{e\beta} \rangle$, also called the diffusion matrix (this reduces to the diffusion coefficient in one dimension of space).

7. Analysis of Multiple-scale Systems

Interfacing ecological and socioeconomic systems presents fundamental challenges because the dominant scales on which these systems operate are so different. Indeed, the problems associated with multiple scale systems are not unique to the interface: Any ecological system (Levin 1992) or any socioeconomic system involves the interplay of processes operating on diverse scales of space, time and organization. Individuals are organized into families, families into populations, populations into species, species into communities, and communities into global assemblages. Small-scale processes, operating typically over relatively fast time scales, become integrated into larger scale processes, operating over longer time scales. Events in ecological time are shaped by evolutionary processes, and in turn drive new episodes of evolution.

These problems are not unique to these systems; scaling problems are among the most pervasive in the physical sciences as well. General circulation models, which are used to predict climate change and other phenomena, require translating small-scale processes into global patterns of circulation, and in turn using those models to predict changes at local scales. This leads to the need for multigrid methods, with varying degrees of resolution at different spatial (and temporal scales). Temporally, the problem of multiple scales also represents an active area of research. If the scales are sufficiently distinct, so that one may be regarded as slow enough that its variables are constant on the fast scale, numerous methods exist, such as singular perturbation theory, to deal with the systems. Guckenheimer (2000) identifies three complementary approaches to such problems: nonstandard analysis, "classical" asymptotic methods, and geometric singular perturbation theory. The development of numerical methods to explore such phenomena is an active area of research.

When the scales in question are very different, substantial simplification is possible by treating the variables on the slow scale as constants for the purpose of the fast-time-scale dynamics, whose asymptotic behavior is then assumed to track the movement of the slow variables on the slow time scale. In enzyme kinetics, this "quasi-steady-state" assumption allows one to assume that enzyme-substrate complexes reach an "equilibrium" concentration, determined by the free concentrations of enzyme and substrate, thereby reducing the dimensionality of the system on the fast time scale. Formally, the mathematical justification for such assumptions are to be found in theorems like those of Tykhonov (1952), but intuitively the idea is clear: On the fast time scale, the slow variables are treated as constants, while on the slow time scale, the fast variables are assumed to be in equilibrium. Through such an assumption, for example, the classical Lotka-Volterra predation equations (with damping) collapse to a logistic equation for the predator if it is assumed that resource dynamics are much faster than predator dynamics.

Formally, the methods of the last section should be replaced by singular perturbation methods, which account explicitly for the relative strength of time scales. Treatment of singular perturbation theory is beyond the scope of this paper, but the reader is referred to the excellent discussion in Lin and Segel (1974), in which the enzyme kinetic problem is treated explicitly.

Multiple-scale methods have been very influential in economics as well, especially since the seminal paper of Simon and Ando (1961). Simon and Ando address the problem of aggregation into sectors, introducing formal notions of system decomposability. In particular, they focus on "nearly decomposable" systems. Based on the network of interactions among agents or firms, a system may be decomposed into clusters within which there is strong interaction, and among which interactions are weak. When this is done, one treats within-cluster dynamics as operating on fast time scales, which reach quasi-steady-states that then interact on the slower time scale. Fisher and Ando (1971) extend this work further, dealing with "completely decomposable" systems, whose internal dynamics may be treated as autonomous on the fast time scale.

Of course, there are problems with this approach, not the least of which is the validity of the assumption of separation of scales, and the notion that clusters reach steady-states on the fast time scale. When the asymptotic dynamics are more complicated, for example chaotic, these simple results do not apply (but see Earn et al. 2000). Second, the approach is essentially linear; more generally, the network of connections will not exhibit constant strengths, and methods are poor for dealing with systems of changing connectivities. Yet the problem of aggregation and dimensional reduction remains a central one, and a rich area of research (Ijiri 1971; Iwasa et al. 1987, 1989; Li and Rabitz 1991a, b).

When scales are separable, as in the earlier examples, the analysis on the slower time scale still can introduce complexities. As the slow variables change, the fast equilibrium (if it exists) may change slowly for a while, but ultimately reach thresholds (bifurcation points) at which qualitative changes take place in the phase portrait. This may result in a discontinuous jump from one equilibrium point to another, a transition to periodic or chaotic behavior, or other transitions. Technically the subject of bifurcation theory, such phenomena also fueled interest thirty years ago in catastrophe theory (Thom 1983). Catastrophe theory attempts to catalogue the fundamental kinds of transitions that can take place. Because catastrophe theory usually did not deal with systems with known dynamics, but tried to intuit underlying dynamics from observed macroscopic behavior, its application generated considerable resistance, and is today much less in evidence outside pure mathematics (Arnol'd 1986). Bifurcation theory, however, which begins from the underlying dynamics and explores the consequences for macroscopic behavior, is a very powerful and popular tool.

The simplest form of bifurcation in Thom's generic scheme is the cusp catastrophe, in which one equilibrium loses stability at a critical threshold in the slow variables, and the system jumps to a new equilibrium. Epidemic systems may fit this scheme neatly, as the level of susceptibility in the population reaches a threshold level at which outbreaks can be sustained. Ludwig et al. (1978) exploited this structure in their consideration of the dynamics of the economically important pest, the spruce budworm. In their model, when the budworm is at low densities, forest quality improves on a slow time scale. When forest quality has reached a sufficiently high level, the budworm population outbreaks, leaping from its low endemic level to a new epidemic one. Now, with the budworm at epidemic levels, forest quality gradually declines until it reaches a point at which the budworm

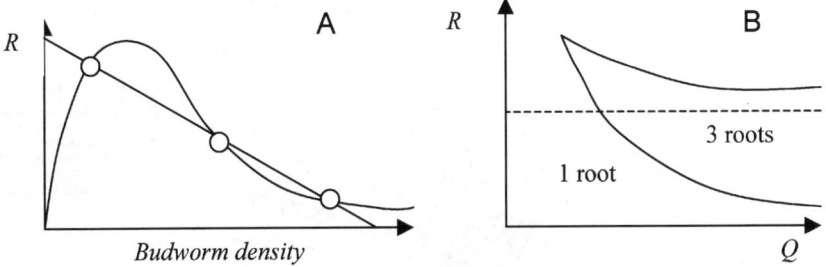

Figure 7. Spruce budworm and cusp bifurcation. A: equilibria of the spruce budworm-forest interaction (circles). The central circle is unstable, while the two other equilibria are stable. R is the reproductive rate of the budworm. B: cusp bifurcation. Depending on the slope and on the position of the straight line in panel A, one or three equilibria can be found. The curve separating these two regimes is cusp-shaped. Adapted from Ludwig et al. (1978), with permission.

can no longer be sustained; the budworm then crashes back to endemic levels (Figure 7). This is an example of a relaxation oscillation. The threshold for explosion is not the same as the threshold for collapse; between the two, the system can exist in either of two quasi-steady-states, depending on history. Thus, it exhibits hysteresis.

The cusp catastrophe has also enjoyed recent influence in the field of environmental economics, through attention to the dynamics of lake systems, especially shallow lakes, subject to eutrophication (e.g., Carpenter et al. 1999; Arrow et al. 2000). In the shallow lake example, the level of phosphorus is a balance between processes that inject the element into the lake (loading), and those that clear it (purification). For intermediate levels of purification, as phosphorus loading is increased, the system may jump from a low (oligotrophic) phosphorus level to a high (eutrophic) one, which may be refractory to amelioration.

The luxury of easily separable scales does not apply to all systems; obviously, when we cannot so separate scales, the dynamics become even more complicated. For example, many systems have intrinsic periodic dynamics, with their own natural scale of fluctuation. When such systems, for example host-parasite systems, are forced say by annual periodicities, things are relatively simple if the scales are very different- one fluctuation will be superimposed on top of another. When the scales are not too different, however, the periodicities interfere with one another, and may give rise to chaos. Chaos, of course, can also arise in systems that are characterized by multiple endogenous periodicities, as when turbulence arises in fluid dynamics, or chaos in population dynamics (May 1974).

So far in this section, most of the focus has been on time scales. The discussion of Simon and Ando's work, however, makes clear that temporal and organizational scales interact, as do spatial and temporal ones (Figure 8). It is well beyond the scope of this paper to discuss the problem of pattern formation in spatial systems, and the interplay between space and time (but see Levin 1979, 1992). In general,

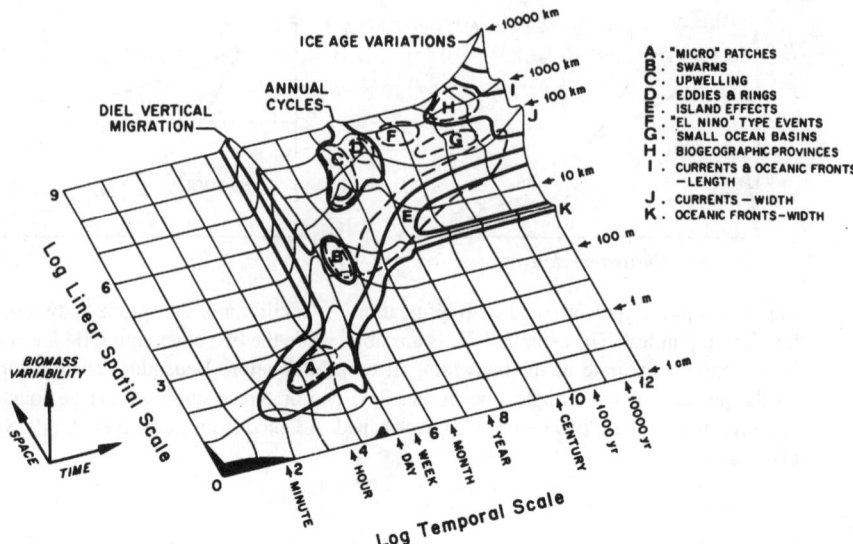

Figure 8. Zooplankton biomass variability across scales (Stommel diagram). Processes are mostly observed along a straight line of increasing spatial and temporal scales. After Haury et al. (1978), reprinted in Levin (1992), with permission.

however, pattern formation emerges from the interplay between processes that break symmetry, and those that stabilize non-uniform structures. Alan Turing, in a classic paper (Turing 1952) motivated by problems in developmental biology, showed how local interactions could give rise to global patterns, through the interplay of activation and inhibition. Turing hypothesized the existence of two chemicals, termed morphogens, one an activator and the other an inhibitor. In his scheme, both chemicals diffuse, but the inhibitor does so on a much broader scale. It is possible then to show that local (short-range) activation breaks symmetry, and long-range inhibition stabilizes it (Segel and Levin 1976; Levin and Segel 1985). Pattern arises, then, through the interplay between processes operating on very different scales. In general, the problem of pattern formation, and the interplay between activation and inhibition, is of deep importance for understanding the organization of societies and economies.

8. Synthesis: Scale and Complex Adaptive Systems

The problem of scale is by no means a purely conceptual one. Careful analyses of the scaling properties of biological systems have had far-reaching consequences in the theory of evolution. Similar reflections have led to major achievements in physics, for example in the study of turbulence, and in unveiling the universal properties of systems near critical transitions. New techniques have emerged for detecting scales (wavelet transforms), and others have exploited the consequences of scale invariance (renormalization group methods).

It is obviously easier to derive robust experimental results with small systems than with large ones. However, problems such as the stability of the climate of the Earth cannot be addressed at the regional scale, and even the current globally coupled ocean-atmosphere-biosphere models can only describe the climate of the Earth on a resolution of half a degree (∼55 kms). They capture many of the large-scale feedbacks, such as the El Nino oscillation, otherwise inaccessible. However, the incorporation subgrid scale processes – what is going on inside the 0.5×0.5 degree cells – is of fundamental importance.

However we have here made a more ambitious proposal. We have suggested that scaling concepts offer an avenue to study heterogeneous assemblies for which the microscopic processes are not known, and probably not knowable (e.g., in the case of social systems), except in terms of their statistical properties. These systems are complex adaptive systems (Levin 2000), which must be studied across levels of organization too.

Acknowledgements

We are pleased to acknowledge the support of the Andrew W. Mellon Foundation and of the David and Lucile Packard Foundation (grant 99-8307), and the stimulus of the Santa Fe Institute, and of the Beijer Institute.

References

Amaral, L., S. Buldyrev, S. Havlin, H. Leschhon, P. Maas, H. E. Stanley and M. Stanley (1997), 'Scaling Behavior in Economics. I. Empirical Results for Company Growth', *Journal de Physique* **7**, 621–633.
Andow, D. A., P. M. Kareiva, S. A. Levin and A. Okubo (1990), 'Spread of Invading Organisms', *Landscape Ecology* **4**, 177–188.
Arnol'd, V. I. (1986), *Catastrophe Theory*. Berlin: Springer-Verlag, 108 pp.
Arrhenius, O. (1921), 'Species and Area', *Journal of Ecology* **9**, 95–99.
Arrow, K., G. Daily, P. Dasgupta, S. Levin, K.-G. Maler, E. Maskin, D. Starrett, T. Sterner and T. Toetenberg (2000), 'Managing Ecosystem Resources', *Environmental Science and Technology* **34**, 1401–1406. Website http://dx.doi.org/10.1021/ES990672T.
Bak, P. and K. Sneppen (1993), 'Punctuated Equilibrium and Criticality in a Simple Model of Evolution', *Physical Review Letters* **71**, 4083–4086.
Bak, P., C. Tang and K. Wiesenfeld (1988), 'Self-organized Criticality', *Physical Review A* **38**, 364–374.
Barabási, A.-L., S. V. Buldyrev, H. E. Stanley and B. Suki (1996), 'Avalanches in the Lung: A Statistical Mechanical Model', *Physical Review Letters* **76**, 2192–2195.
Batchelor, G. K. (1953). *The Theory of Homogeneous Turbulence*. Cambridge University Press.
Binney, J. J., N. J. Dowrick, A. J. Fisher and M. E. J. Newman (1992), *The Theory of Critical Phenomena – An Introduction to the Renormalization Group*. Oxford: Oxford Science Publications, Clarendon Press, 464 pp.
Bogoliubov, N. N. (1946), *Studies in Statistical Mechanics*, in J. de Boer and G. E. Uhlenbeck, eds. Amsterdam: North-Holland Publishing Company.
Bogoliubov, N. N. and D. V. Shirkov (1959), *Introduction to the Theory of Quantized Fields*. New York: Interscience Publishers, 720 pp.

Bouchaud, J.-P. and M. Potters (2000), *Theory of Financial Risks*. Cambridge University Press.
Broadbent, S. R. and J. M. Hammersley (1957), 'Percolation Processes', *Proceedings of the Cambridge Philosophical Society* **53**, 629–645.
Brock (1999), 'Scaling in Economics: A Reader's Guide', *Industrial and Corporate Change* **8**, 409–446.
Brown, J. H. and M. V. Lomolino (1998), *Biogeography*. Sinauer: Sunderland Mass.
Buttel, L., R. Durrett and S. Levin (2002), 'Competition and Species Packing in Patchy Environments', *Theoretical Population Biology* **61**, 265–276.
Calder III, W. A. (1984), *Size, Function, and Life History*. Cambridge Mass: Harvard University Press.
Carpenter, S. R., D. Ludwig and W. A. Brock (1999), 'Management of Eutrophication for Lakes Subject to Potentially Irreversible Change', *Ecological Applications* **9**, 751–771.
Chave, J., M. Dubios and B. Riéra (2001), 'Estimation of Biomass in a Neotropical Forest of French Guiana: Spatial and Temporal Variability', *Journal of Tropical Ecology* **17**, 79–96.
Chave, J., H. Muller-Landau and S. Levin (2002), 'Comparing Classical Community Models: Theoretical Consequences for Patterns of Diversity', *American Naturalist* **159**, 1–23.
Clark, J. S., C. Fastie, G. Hurtt, S. T. Jackson, C. Johnson, G. King and M. Lewis (1998), 'Reid's Paradox of Rapid Plant Migration', *BioScience* **48**, 13–24.
Clark, J. S. (1998), 'Why Trees Migrate So Fast: Confronting Theory with Dispersal Biology and the Paleorecord', *American Naturalist* **152**, 204–224.
Connor, E. F. and E. D. McCoy (1979), 'The Statistics and Biology of the Species-area Relationship', *American Naturalist* **113**, 791–833.
Crawley, M. J. and J. E. Harral (2001), 'Scale Dependence in Plant Biodiversity', *Science* **291**, 864–868.
Durlauf, S. (1993), 'Nonergodic Economic Growth', *Review of Economic Studies* **60**, 349–366.
Durrett, R. and S. Levin (1994), 'Stochastic Spatial Models: A User's Guide to Ecological Applications', *Philosophical Transactions of the Royal Society of London B* **343**, 329–350.
Durrett, R. and S. A. Levin (1996), 'Spatial Models for Species-area Curves', *Journal of Theoretical Biology* **179**, 119–127.
Earn, D., S. Levin and P. Rohani (2000), 'Coherence and Conservation', *Science* **290**, 1360–1363.
Enquist, B. J., J. H. Brown and G. H. West (1998), 'Allometric Scaling of Plant Energentics and Population Density', *Nature* **395**, 163–165.
Ferziger, J. H. and H. G. Kaper (1972), *Mathematical Theory of Transport Processes in Gas*. Amsterdam, London: North-Holland Publishing Company, 579 pp.
Fisher, F. M. and A. Ando (1971), 'Two Theorems on "Ceteris Paribus" in the Analysis of Dynamic Systems', in H. M. Blalock Jr., ed., *Causal Models in the Social Sciences* (pp. 190–199). Chicago: Aldine Publishing Co.
Flierl, G., D. Grünbaum, S. Levin and D. Olson (1999), 'From Individuals to Aggregations: The Interplay between Behavior and Physics', *Journal of Theoretical Biology* **196**, 397–454.
Fortuin, C. M. and P. W. Kastelyn (1972), 'On the Random Cluster Model, I. Introduction and Relation to Other Models', *Physica* **57**, 536–564.
Fragoso, J. (1997), 'Tapir-generated Seed Shadows: Scale-dependent Patchiness', *Journal of Ecology* **85**, 519–529.
Frisch, U. (1995), *Turbulence: The Legacy of A.N. Kolmogorov*. Cambridge: Cambridge University Press, 296 pp.
Gell-Mann, M. (1994), *The Quark and the Jaguar*. New York: Freeman and Co.
Gentry, A. H. (1992), 'Tropical Forest Diversity: Distributional Patterns and Their Conservational Significance', *Oikos* **63**, 19–28.
Grassberger, P. and A. de la Torre (1979), Reggeon Field Theory (Schlogl's First Model) on a Lattice: Monte-Carlo Calculations of Critical Behaviour', *Annals of Physics* **122**, 373.

Guckenheimer, J. (2000), Numerical computation of canards. Website http://www.cam.cornell.edu/guckenheimer/timescale.html.

Harris, T. E. (1974), 'Contact Interactions on a Lattice', *Annals of Probability* **2**, 969.

Harte, J., A. Kinzig and J. Green (1999), 'Self-similarity in the Abundance of Species', *Science* **284**, 334–336.

Hastings, H. M. and G. Sugihara (1993), *Fractals. A User's Guide for the Natural Sciences*. Oxford: Oxford Science Publications, Oxford University Press, 235 pp.

Haury, L. R., J. A. McGowan and P. H. Wiebe (1978), 'Patterns and Processes in the Time-space Scales of Plankton Distributions', in J. H. Steele, ed., *Spatial Pattern in Plankton Communities* (pp. 277–327). New York, U.S.A.: Plenum.

Hinrichsen, H. (2000), 'Nonequilibrium Critical Phenomena and Phase Transitions into Absorbing States', *Advances of Physics* **49**, 815. Available online at http://xxx.lanl.gov (Los Alamos e-archive), section Condensed Matter, number cond-mat/0001070, 153 pp.

Horn, H. S. (2000), 'Twigs, Trees, and the Dynamics of Carbon in the Landscape', in J. H. Brown and G. B. West, eds., *Scaling in Biology* (pp. 199–220). Santa Fe Institute, Oxford University Press, 352 pp.

Hubbell, S. P. (1997), 'A Unified Theory of Biogeography and Relative Species Abundance and Its Application to Tropical Rain Forests and Coreal Reefs', *Coral Reefs* **16**, S9–S21.

Ijiri, Y. (1971), 'Fundamental Queries in Aggregation Theory', *Journal of the American Statistical Association* **66**, 766–782.

Ijiri, Y. and H. Simon (1977), *Skew Distributions and the Size of Business Firms*. Amsterdam: North Holland.

Iwasa, Y., S. A. Levin and V. Andreasen (1987), 'Aggregation in Model Ecosystems. I. Perfect Aggregation', *Ecological Modelling* **37**, 287–302.

Iwasa, Y., S. A. Levin and V. Andreasen (1989), 'Aggregation in Model Ecosystems. II. Approximate Aggregation', *IMA Journal of Mathematics Applied in Medicine* **6**, 1–23.

Jacob, F. (1977), 'Evolution and Tinkering', *Science* **196**, 1161–1166.

Kadanoff, L. P. (1966), 'Scaling Laws for Ising Models near T_c', *Physics* **2**, 263.

Kauffman, S. (1993), *The Origins of Order*. Oxford: Oxford University Press.

Kesten, H. (1980), 'The Critical Probability for Bond Percolation on the Square Lattice Equals 1/2', *Communications in Mathematical Physics* **74**, 41–59.

Kinzig, A. P., S. A. Levin, J. Dushoff and S. W. Pacala (1999), 'Limiting Similarity, Species Packing, and System Stability for Hierarchical Competition-colonization Models', *American Naturalist* **153**, 371–383.

Kolmogorov, A. N. (1941), 'The Local Structure of Turbulence in Incompressible Viscous Fluid for Very Large Reynolds Number', *Comptes Rendus Académie des Sciences URSS* **30**, 301.

Levin, S. A. (1979), 'Non-uniform Stable Solutions to Reaction-diffusion Equations: Applications to Ecological Pattern Formation', in H. Haken, ed., *Pattern Formation by Dynamic Systems and Pattern Recognition* (pp. 210–222). Berlin: Springer-Verlag.

Levin, S. A. and L. A. Segel (1985), 'Pattern Generation in Space and Aspect', *SIAM Review* **27**, 45–67.

Levin, S. A. (1991), 'The Problem of Relevant Detail', in S. Busenberg and M. Martelli, eds., *Differential Equations. Models in Biology, Ecology, and Epidemiology* (pp. 9–15). Springer Verlag.

Levin, S. A. (1992), 'The Problem of Pattern and Scale in Ecology', *Ecology* **73**, 1943–1967.

Levin, S. A. (1993), 'Concept of Scale at the Local Level', in J. R. Ehlringer and C. B. Field, eds., *Scaling Physiological Processes. Leaf to Globe* (pp. 7–19). Academic Press, 388 pp.

Levin, S. A. (2000), 'Complex Adaptive Systems: Exploring the Known, the Unknown and the Unknowable', in F. Browder, ed., *Mathematical Challenges of the XXIth Century*. American Mathematical Society, Providence RI.

Levin, S. A., A. Morin and T. H. Powell (1989), 'Patterns and Processes in the Distribution and Dynamics of Antartic Krill', in *Scientific Committee for the Conservation of Antarctic Marine Living Resources – Selected Scientific Papers, Part 1, SC-CAMLR-SSp/5* (pp. 281–299). Australia: Hobart, Tasmania.

Levin, S. A. and S. W. Pacala (1997), 'Theories of Simplification and Scaling of Spatially Distributed Processes', in D. Tilman and P. Kareiva, eds., *Spatial Ecology: The Role of Space in Population Dynamics and Interspecific Interactions* (pp. 271–296). Princeton, NJ: Princeton University Press.

Levin, S. A. (1999), *Fragile Dominion*. Massachussets: Helix Books, Perseus Reading.

Levins, R. (1968), *Evolution in Changing Environments*. Princeton, NJ: Princeton University Press.

Li, G. and H. Rabitz (1991a), 'A General Lumping Analysis of a Reaction System Coupled with Diffusion', *Chemical Engineering Science* **46**, 2041.

Li, G. and H. Rabitz (1991b), 'A General Analysis of Lumping', in G. Astarita and S. I. Sandler, eds., *Chemical Kinetics, in Kinetic and Thermodynamic Lumping of Multicomponent Mixtures* (pp. 49–62). Amsterdam: Elsevier Science Publishers B.V.

Lin, C. C. and L. A. Segel (1974), *Mathematics Applied to Deterministic Problems in the Natural Sciences*. New York: Macmillan.

Ludwig, D. and C. J. Walters (1985), 'Are Age-structured Models Appropriate for Catch-effort Data?', *Canadian Journal of Fish Aquatic Science* **42**, 1066–1072.

Ludwig, D., D. D. Jones and C. S. Holling (1978), 'Qualitative Analysis of Insect Outbreak Systems: The Spruce Budworm and Forest', *Journal of Animal Ecology* **47**, 315–332.

MacAuley, F. R. (1922), 'Pareto's Laws and the General Problem of Mathematically Describing the Frequency of Income', *Income in the United States: Its Amount ands Distribution 1909–1919*. New York: National Bureau of Economic Research.

Major, J. Endemism (a botanical perspective), In A. A. Myers and P. S. Giller, eds., *Analytical Biogeography – An Integrated Approach to the Study of Animal and Plant Distributions* (pp. 117–146). London: Chapman and Hall.

Makse, H. A., S. Havlin and H. E. Stanley (1995), 'Modelling Urban Growth Patterns', *Nature* **377**, 608–612.

Malamud, B. D., G. Morein and D. L. Turcotte (1998), 'Forest Fires: An Example of Self-organized Critical Behavior', *Science* **281**, 1840–1842.

Mandelbrot, B. B. (1963), 'New Methods in Statistical Economics', *Journal of Political Economy* **71**, 421–440.

Mandelbrot, B. B. (1975), *Les Objets Fractals. Forme, Hasard et Dimension*. Flammarion, Paris: Nouvelle Bibliothèque Scientifique, 190 pp.

Mandelbrot, B. B. (1997), *Fractals and Scaling in Finance*. Springer-Verlag.

May, R. M. (1974), 'Biological Populations with Non-overlapping Generations: Stable Points, Stable Cycles, and Chaos', *Journal of Theoretical Biology* **49**, 511–524.

May, R. M. and P. F. Stumpf (2000), 'Species-area Relationships for Tropical Forests', *Science* **290**, 2084–2086.

McMahon, T. A. (1975), 'The Mechanical Design of Trees', *Scientific American* **233**, 93–102.

Mills, T. C. (1999), *The Econometric Modelling of Financial Time Series*. Cambridge University Press, 372 pp.

Mollison, D. and S. A. Levin (1995), 'Spatial Dynamics of Parasitism', in B. T. Greenfell and A. P. Dobson, eds., *Ecology of Infectious Diseases in Natural Populations* (pp. 384–398). Cambridge, UK: Cambridge University Press.

Nathan, R., U. N. Safriel, I. Noy-Meir and G. Schiller (2000), 'Spatiotemporal Variation in Seed Dispersal and Recruitment near and Far From *Pinus Halepensis* Trees', *Ecology* **81**, 2156–2161.

Nathan, R. (2001), 'Dispersal Biogeography', in S. Levin, ed., *Encyclopedia of Biodiversity* (pp. 127–152). Academic Press.

Niklas, K. (1992), *Plant Biomechanics: An Engineering Approach to Plant Form and Function.* Chicago and London: University of Chicago Press, 607 pp.

O'Brien, S. T., S. P. Hubbell, P. Spiro, R. Condit and R. B. Foster (1995), 'Diameter, Height, Crown and Age Relationships in Eight Neotropical Tree Species', *Ecology* **76**, 1926–1939.

Okubo, A. and S. A. Levin (1989), 'A Theoretical Framework for Data Analysis of Wind Dispersal of Seeds and Pollen. *Ecology* **70**, 329–338.

Pareto, V. (1896), *Cours d'Economie Politique.* Reprinted in *Pareto. Oeuvres Complètes.* Droz, Geneva, 1965.

Peters, R. H. (1983), *The Ecological Implications of Body Size.* Cambridge: Cambridge University Press, 324 pp.

Pitman N. C. A., J. Terborgh, M. R. Silman and P. V. Nuñez (1999), 'Tree Species Distribution in an Upper Amazonian Forest', *Ecology* **80**, 2651–2661.

Plotkin J. B., M. D. Potts, D. W. Yu, S. Bunyavejchewin, R. Condit, R. Foster, S. Hubbell, J. LaFrankie, N. Manokaran, L. Hua Seng, R. Sukumar, M. A. Nowak and P. S. Ashton (2000), 'Predicting Species Diversity in Tropical Forests', *Proceedings of the National Academy of Sciences U.S.A.* **97**, 10850–10854.

Plotkin, J. B., J. Chave and P. S. Ashton (2002), 'Cluster Analysis of Spatial Patterns in Malaysian Tree Species', *American Naturalist* **60**, 629–644.

Prance, G. T. (1973), 'Phytogeographic Support for the Theory of Pleistocene Forest Refuges in the Amazon Basin, Based on Evidence from Distribution Patterns in Caryocaraceae, Chrysobalanaceae, Dichapetalaceae and Lecythidaceae', *Acta Amazônica* **3**, 5–28.

Prance, G. T. (1978), 'Floristic Inventory of the Tropics: Where Do We Stand?', *Annals of the Missouri Botanical Garden* **64**, 659–684.

Prance, G. T. (1994), 'A Comparison of the Efficacy of Higher Taxa and Species Numbers in the Assessment of Biodiversity in the Tropics', *Philosophical Transactions of the Royal Society of London B* **345**, 89–99.

Reynolds, P. J., W. Klein and H. E. Stanley (1977), 'A Real-space Renormalization Group for Site and Bond Percolation', *Journal of Physics C* **10**, L167.

Rodríguez-Iturbe, I. and A. Rinaldo (1997), *Fractal River Basins. Chance and Self-Organization.* Cambridge: Cambridge University Press, 547 pp.

Schmidt-Nielsen, K. (1984), *Scaling. Why is Animal Size so Important?* Cambridge: Cambridge University Press, 241 pp.

Segel, L. A. and S. A. Levin (1976), 'Applications of Nonlinear Stability Theory to the Study of the Effects of Diffusion on Predator Prey Interactions', in R. A. Piccirelli, ed., *Topics in Statistical Mechanics and Biophysics: A Memorial to Julius L. Jackson.* AIP New York.

Shinozaki, K., K. Yoda, K. Hozumi and T. Kira (1964), 'A Quantitative Analysis of Plant Form – The Pipe Model Theory I. Basic Analyses', *Japanese Journal of Ecology* **14**, 97–105.

Simon, H. A. and A. Ando (1961), 'Aggregation of Variables in Dynamic Systems', *Econometrica* **29**, 111–138.

Skellam, J. G. (1951), 'Random Dispersal in Theoretical Populations', *Biometrika* **38**, 196–218.

Sornette, A. and D. Sornette (1989), 'Self-organized Criticality and Earthquakes', *Europhysics Letters* **9**, 197–202.

Stanley, H. E. (1971), *Introduction to Phase Transitions and Critical Phenomena.* Oxford: Oxford Science Publications, Clarendon Press.

Stull, R. B. (1988), *An Introduction to Boundary Layer Meteorology.* Dordrecht, Boston, London: Atmospheric Sciences Library, Kluwer Academic Publishers, 670 pp.

Thom, R. (1983), *Mathematical Models of Morphogenesis.* New York: Halsted Press, 305 pp.

Turing, A. (1952), 'The Chemical Basis of Morphogenesis', *Philosophical Transactions of the Royal Society of London* B **237**, 37–72.

Tykhonov, A. N. (1952), *Systems of Differential Equations Containing Small Parameters Multiplying the Derivatives.* Mat.

Vogel, S. (1994), *Life in Moving Fluids*. Princeton, NJ: Princeton University Press, 467 pp.
West, G. B., J. H. Brown and B. J. Enquist (1997), 'A General Model for the Origin of Allometric Scaling Laws in Biology', *Science* **276**, 122–126.
Whittaker, R. J., M. B. Bush and K. Richards (1989), 'Plant Recolonization and Vegetation Succession on the Krakatau Islands, Indonesia', *Ecological Monographs* **59**, 59–123.
Williams, C. B. (1964), *Patterns in the Balance of Nature and Related Problems in Quantitative Ecology*. New York: Academic Press.
Wilson, K. G. and J. Kogut (1974), 'The Renormalization Group and the ε-expansion', *Physics Reports* **12C**, 75–200.
Yeomans, J. M. (1992), *Statistical Mechanics of Phase Transitions*. Oxford: Oxford Science Publications, Clarendon Press, 153 pp.
Zipf, G. K. (1949), *Human Behavior and the Principle of Least Effort*. Addison Wesley.

Appendix

Now, let us consider the macroscopic scale, which is experimentally accessible. Practically, one is interested in the density of particles in a domain of size dX around a location X, and within a range of velocity dV around V, $n(X, V)dXdV$. We use capital letters to avoid possible confusion between the microscopic variables (x_i, v_i) and the macroscopic ones (X, V). The rate equation for $n(X, V)$ is:

$$n(X', V', t + \delta t) = \int_\Omega P(X', V'|X, V; \delta X_e, \delta V_e) n(X, V, t) dX dV \qquad (2)$$

where $P(X', V'|X, V; \delta X_e, \delta V_e)$ is the transition rate from the elementary volume around (X, V) in the phase space to the elementary volume around (X', V'). This equation describes the dynamics of the density $n(X, V)$. At this point, we are dealing with a macroscopic equation: (X, V) is not the position of a particle, but that of a small volume of "fluid" made out of a large number of particles. Well-known methods exist to transform equation (3) into a tractable dynamic equation. The simplest form is Boltzmann's equation. Assuming that the collision term $(\delta x_e, \delta v_e)$ is negligible in equation (1), the microscopic description coincides with the macroscopic one, and we rewrite (1) as:

$$n(X + V\delta t, V + A\delta t, t + \delta t) = n(X, V, t)$$

where A is the acceleration (force per unit mass) of the element of system. This is equivalently rewritten $P(X', V'|X, V; \delta X_e, \delta V_e) = P(X + V\delta t, V + A\delta t|X, V)$. Taking the limit $\delta t \to 0$ yields the following differential equation

$$\frac{\partial n(X, V, t)}{\partial t} + \sum_\alpha V_\alpha \frac{\partial n(X, V, t)}{\partial X_\alpha} + \sum_\alpha A_\alpha \frac{\partial n(X, V, t)}{\partial V_\alpha} = 0 \qquad (3)$$

where $X = \{X_1, X_2, X_3\}$. Now, if collisions are present, we can generalize equation (3) by:

$$\frac{\partial n(X, V, t)}{\partial t} + \sum_\alpha V_\alpha \frac{\partial n(X, V, t)}{\partial X_\alpha} + \sum_\alpha A_\alpha \frac{\partial n(X, V, t)}{\partial V_\alpha} = \frac{\partial n(X, V, t)}{\partial t}\bigg|_{coll}$$

This is Boltzmann's equation. Of course, the interesting part of this equation is the right-hand term, and one must deduce it from the microscopic laws (1) and from the conservation of momentum ($m_1 v_1 + m_2 v_2 = cte$) and of kinetic energy ($m_1 v_1^2 + m_2 v_2^2 = cte$) during a collision. This can be carried out using macroscopic arguments, as in the Chapman-Enskog derivation (Chandrasekar 1943). A partial differential equation, the Fokker-Planck equation, can be used as an alternative kinetic equation in certain limiting cases:

$$\frac{\partial n(X, V, t)}{\partial t} + \left\{ \sum_\alpha \frac{\partial}{\partial X_\alpha} V_\alpha + \sum_\alpha \frac{\partial}{\partial V_\alpha} A_\alpha \right\} n(X, V, t) \qquad (4)$$

$$= \frac{1}{2} \left\{ \sum_{\alpha\beta} \frac{\partial}{\partial V_\alpha \partial V_\beta} \Gamma_{\alpha\beta} \right\} n(X, V, t)$$

Equation (4) is valid in the absence of random hops ($\delta X_e = 0$). In this case, the matrix $\Gamma_{\alpha\beta}$ is the covariance matrix $\langle \delta V_{e\alpha} \delta V_{e\beta} \rangle$. For a discussion of this approach for natural systems, see Flierl et al. (1999) and Levin (2001). The above procedure can also be directly derived from the microscopic theory by constructing the Liouville equation, a microscopic equivalent of equation (3), in which collisions are added. This procedure is due to Bogoliubov (1946), and often referred to as the BBGKY procedure (after its authors, Bogoliubov, Born, Green, Kirkwood, and Yvon; see Ferziger and Kaper 1972).

Convex Relationships in Ecosystems Containing Mixtures of Trees and Grass

R.J. SCHOLES
CSIR Division of Water, Environment and Forest Technology, P.O. Box 395, Pretoria 0001, South Africa

Abstract. The relationship between grass production and the quantity of trees in mixed tree-grass ecosystems (savannas) is convex for all or most of its range. In other words, the grass production declines more steeply per unit increase in tree quantity at low tree cover than at high tree cover. Since much of the economic value in savannas is ultimately derived from grass, and the main mechanism controlling the tree-grass balance is dependent on the production of grassy fuel for fires, this non-linearity has the effect of creating two savanna configurations. One has a low tree density and supports a viable grazing enterprise, while the other has dense tree cover and a frequently non-viable grazing enterprise. The non-linearity is suggested here to have two main sources: the geometry of the spatial interaction between tree root system and grasses, and the effect of differing phenology (the time course of leaf area exposure) on the acquisition of water and nutrients. The existence of the non-linearity reduces the resilience of the generally-preferred "open" configuration, and increases the resilience of the less-desirable "closed" configuration.

Key words: competition, primary productivity, resilience, savannas

1. Introduction

A large fraction of the Earth's land surface is covered by ecosystems in which the plant production is generated from mixtures of trees and grasses. The most extensive case is the approximately one-eighth of the surface occupied by tropical savannas (Scholes and Hall 1996). Similar ecosystems occur outside the tropics as well, and in certain cultivated systems.

The key issue in the management of these systems for human benefit relates to the proportions of the mixture, since more tree means less grass. Where the principal benefits are based on the grass component, such as in grazing systems, the strategy seems obvious: maximise grass production. In the short term, maximum grass production is typically (but not always: see Figure 1) achieved at zero tree presence, but since removal of the trees is a significant and recurring expense, and may incur other environmental costs, the objective typically involves finding the level of tree presence that strikes a balance between loss of grass production and loss of other ecosystem services.

The quantity of woody plants in the system can be measured in a variety of ways: tree canopy cover percentage, tree biomass, leaf area, stand basal area or

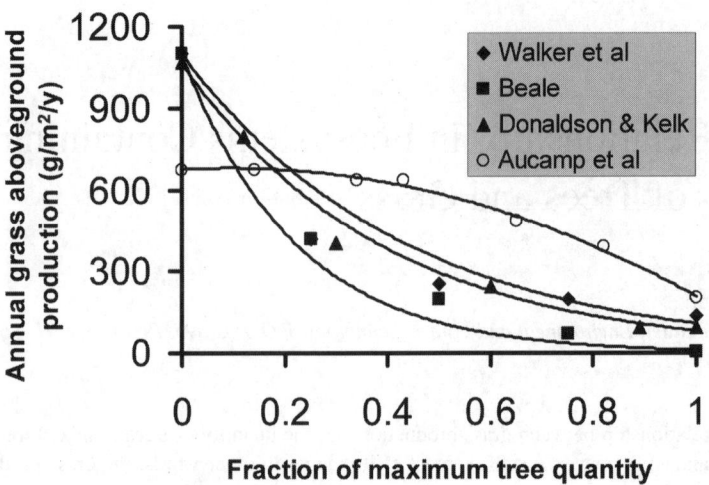

Figure 1. Empirical relationships between the quantity of trees in the system (measured in various ways, and here normalised so that the largest tree quantity reported in the study is equal to 1) and the aboveground grass production.

number of trees per unit area. The last-mentioned is not a good measure, since most systems have many small trees that have little functional impact on grasses. All the other "treeness" measures are highly correlated with one another and with impacts on grass production. Unless otherwise stated, this paper uses stand basal area (i.e., the sum of the cross-sectional area of the stems near the ground level, per unit ground area, m^2/ha) as the index of tree quantity, and it is here given the symbol X. Note that this is a measure of "stock", whereas in general the grasses are measured in terms of production ($g/m^2/y$), which is a "flux".

Grass tillers are short-lived (< 1 year), so the useful and grazable stock is more-or-less equal to the within-growing-season production. The use of this mixed terminology has the potential for causing confusion, particularly in graphical analysis where the state of the system is represented by the stocks of the two components. For the key analysis in this paper, the grass production flux is converted to a stock by accumulating it, less losses due to fire, herbivory and decay.

The trade-off curve between the stand basal area and the grass production is central to both the economic use of savannas and their ecological dynamics. In every recorded case it has a strongly non-linear form. The grass production declines as tree basal area increases. In almost all cases this relationship is convex over its entire range; i.e., the grass production declines more rapidly for the initial increments in tree basal area than it does for subsequent increments (Scholes and Archer 1997). In a small proportion of cases, there is little or no impact on grass production (or even a slight increase) for the first increments of tree basal area above zero, followed by a transition to a inverse convex relationship above a certain level of basal area per hectare (Figure 1).

Savanna systems around the world have been observed to be unstable in terms of the proportions of the mixture, under certain forms of management. In particular, following the introduction of sustained heavy grazing by domestic stock, usually accompanied by reduction in the fire frequency and intensity, the quantity of trees tends to abruptly and apparently irreversibly increase, often to a point where the grass production is so suppressed that pastoralism becomes commercially unviable.

This paper addresses two questions:
- what is the mechanistic basis of the observed convexity in the grass production vs tree basal area curve?
- does the shape of this curve have anything to do with the abrupt transition between different configurations of tree-grass systems?

The analysis presented here is one way of conceptualising savannas. Like all ecosystems, savannas can be viewed through a variety of lenses, whose validity depends on the objectives of the study. The relatively simple and abstracted model presented here aims to explain the widely-observed non-linearity in grass production as a function of tree cover, and to explore its consequences for system dynamics. It does not address species composition dynamics within the grass or tree layers, nor the complexities of grazing behaviour.

2. The Convex Relationship between Tree Basal Area and Grass Production

In the semi-arid to dry subhumid climates that dominate the tropical savanna extent, aboveground grass production, in the absence of tree cover and accumulated over the period of the wet season (i.e., g $DM/m^2/y$, where DM stands for dry matter), has repeatedly been shown to be a linear function of rainfall received during the growing period (equation 1; Rutherford 1980; Scholes 1993). This has been proposed as evidence that grass production in these systems is water-limited; I suggest that it shows that grass production is both water and nutrient limited (Scholes 1997). For instance, fertilisation of these grasslands with nitrogen leads to a relation that is still linear, but with a steeper slope. The simple interpretation is that water availability controls the *duration* of grass growth in these episodically-wetted soils, while nutrient availability determines the *rate of growth* when water is available. The slope of such a linear relationship between Rainfall (R) and aboveground grass production (P_g) is a reflection of the nutrient availability in the system. It is sometimes referred to as the "rain use efficiency" (le Houerou 1984). The intercept of the line, on the other hand, reflects the availability of soil water. For instance, more water is needed to commence production on a clayey soil than on a sandy soil because of the greater initial losses to evaporation and runoff on clayey soils. This is clearer when the equation is written in the form of a slope (a) and an x-axis intercept (c), rather than the more usual y-axis intercept:

$$P_g = a(R - c) \tag{1}$$

A compilation of data from a variety of savanna-climate grasslands provides the following empirical predictors of the slope and x-axis intercept, as functions of the sand content of the soil (s, as a percentage). For soils between 64 and 92% sand content

$$a = -0.0376s + 3.442 \qquad (2)$$

For soils above 92% sand content, a can be given the constant value of 0.1, and below 64%, the constant value of 1.1 g/m^2/y/mm. The value of c is a function of the value of a, since they are both ultimately related to soil texture:

$$c = 328 - \frac{142}{a} \qquad (3)$$

Measurement of tree radial increment growth (the growth revealed by tree rings, or by measuring the increasing circumference of the stem) shows that wood production is related to both rainfall and "site quality", a factor that includes soil depth, landscape position and fertility. Primary production by trees is thus also related to water and possibly to nutrient availability, but in less direct ways than in grasses. This is because trees have a greater storage capacity for water, nutrients and carbohydrates than grasses do, and have a greater rooting volume and depth. Thus in many circumstances they are able to carry over a significant quantity of water, nutrients and carbohydrates between years, as well as conserving and recycling a substantial fraction of their nutrients internally at the time of leaf drop. Thus production by trees in a given year needs to consider rainfall not only in the current growing season, but in the previous one as well.

When trees and grasses grow together, several sometimes opposing interactive mechanisms are operative (Scholes and Archer 1997):
1. Competition for water and nutrients between the root systems of the trees and grass;
2. Reduction of the photosynthetically-active radiation and rainfall reaching the grass canopy, through prior interception by a tree canopy;
3. Improved growing conditions for the grasses immediately below the canopy of a tree due to local nutrient enrichment and improved soil water conservation;
4. A risk of scorching for the trees when the flammable dry grass burns.

Although trees and grasses may be occupying the same land area, their root systems are not necessarily occupying entirely the same volume. The grass cover is usually characterised as "continuous" whereas the tree cover is "discontinuous". Thus, in purely geometrical terms, there may be areas where the grasses are relatively free of tree influence, simply because they are in the interstices of the tree rooting radii (Figure 2).

Since the root reach of savanna trees has been shown to be several multiples of the crown radius (Rutherford 1983; Groot and Soumare 1995), a patch of savanna can be considered to comprise three more-or-less spatially distinct sub-habitats:
1. *the area immediately below the tree crowns*, where the grasses are subject to shading and the interception of water by the crowns, and although exposed to root competition from the trees, generally benefit from the enhanced nutrient status under the crown and a lower evaporative demand;

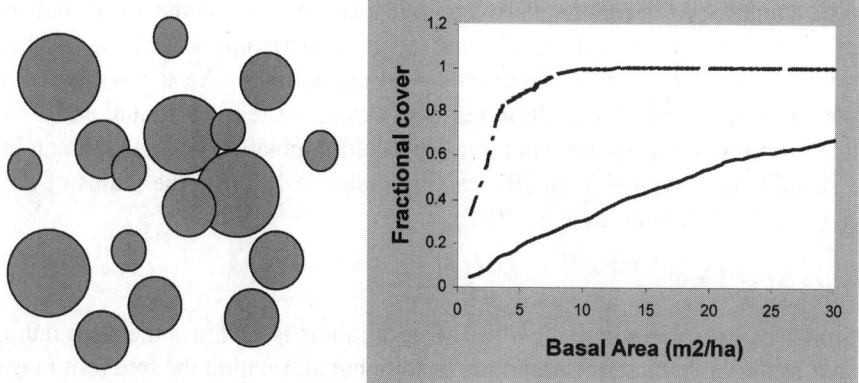

Figure 2. The fraction of the vegetation patch that is beneath tree canopies and between tree canopies can be calculated from knowledge of the tree canopy diameter distribution and the spatial dispersion pattern of trees in the landscape. For simple distributions of both an analytical solution is possible, but for complex distributions, an empirical fit to a numerical solution is more practical. The curves described here assume random spatial distribution, a crown diameter 10 times greater than the stem diameter, and a root diameter 2.5 times the crown diameter.

2. *the area within the reach of tree roots, but beyond the crown*, where shading is minor, but there is competition with trees for water and nutrients; and
3. *the area beyond both the crown and the reach of tree roots*. Grass production here should be the same as grass production in the complete absence of trees. This area is small in savannas with a high tree cover, but can be significant in sparsely-treed savannas.

The aboveground production by grasses (P_g) in a mixed tree-grass situation can therefore be expressed as

$$P_g = P_{g1}A_1 + P_{g2}A_2 + P_{g3}A_3 \qquad (4)$$

where A_1, A_2 and A_3 are the fractions of the patch occupied by the subcanopy, tree root zone and tree root-free subhabitats respectively, and P_{g1}, P_{g2} and P_{g3} are the grass productivities per unit area in each zone.

The area of each subhabitat can be calculated geometrically if the distributions of tree sizes are known as well as their spatial distribution. The tree size distribution in undisturbed, self-regenerating populations follows an "inverse J" distribution, i.e., an exponential decline in the numbers of individuals in each successive size class due to the effect of tree mortality. It may deviate substantially from this ideal form – for instance, many populations show episodic recruitment and mortality, giving multi-modal size distributions. The default assumption regarding spatial distribution is that the location of each tree is random (i.e., independent of every other tree). This assumption is approximated in some naturally-regenerating populations, but not all. Clumping (at several scales, and operating differently for different species in a mixture) is commonly observed, and so is "over-dispersion" (i.e., a tendency towards regularity, with its extreme form in the hexagonal spacing of an orchard). Since both the size and location distributions can be quite complex, this paper calculated a family of empirical predictors of A_1 and A_2, based on the output of a numerical simulation fed with a range of feasible spatial and size-class assumptions. Thus, for a randomly distributed population of trees with an inverse J size distribution, and where the canopy radius is 10 times the radius of the stem (a typical value: Scholes et al. 2002),

$$A_1 = 1 - e^{-0.0379X} \tag{5}$$

Similarly, for trees where the effective root radius is 25 times the stem radius, the area of the subhabitat outside of the canopy but still within the tree root range is

$$A_2 = (1 - e^{-0.4784X}) - A_1 \tag{6}$$

Note that these equations suggest the root reach is about 2.5 times the canopy reach. Researchers who have excavated savanna tree root systems report higher numbers, around 6 times canopy radius (Rutherford 1983; Chin Ong 1996). When these higher values are used in the simulation, effectively no part of the savanna is free of competition with tree roots, which does not accord with field observations. The solution to this apparent contradiction is probably that the roots far from the tree cannot effectively colonise the entire soil volume of the circle described by their most distal reach. The density of roots declines approximately exponentially with distance from the stem (van Noordwijk 1996) Thus an "effective radius" can be described that is smaller than the maximum radius; about 2.5 times greater than crown radius seems to provide satisfactory results. Finally,

$$A_3 = 1 - (A_1 + A_2) \tag{7}$$

In the context of the focus of this paper on non-linearities, the important point is that regardless of the assumptions made about canopy and root dimensions and plant distributions, the fraction of area that grasses have for their exclusive use, and the unshaded area, both decline in a convex form with increasing stand basal area of trees. The degree of convexity increases with dispersed spatial distributions and decreases with clumped distributions.

The grass productivity in subhabitat 3 (where there is no tree root or crown influence) is given by the equation for grass production in the absence of trees (equation 1)

$$P_{g3} = a(R - c) \tag{8}$$

The grass productivity in subhabitat 2 (root competition but no crown influence) depends on the proportion of the available resources in the soil (water and nutrients) that is captured by trees, and is thus unavailable for grasses. The classical hypothesis in this regard is that grasses have preferential access to resources in the topsoil, while trees have preferential access in the subsoil (Walter 1971). Empirical evidence (e.g., Scholes and Walker 1993) indicates that the effective rooting-depth niche separation is minor by comparison to the niche separation in time, largely because water and nutrients are both mostly to be found in the topsoil, and thus trees and grass both have most of their roots there. The niche separation in time results from the fact that trees expand their leaf area more rapidly at the beginning of the growing season, and retain them for longer at the end of the growing season, than grasses do. Even though the tree relative growth rate may be slightly less than that of grasses, and in some cases the grass peak leaf area may exceed the tree leaf area in the middle of the growing season, the beginning-and-end of season windows of near-exclusive use provide a strong advantage to trees.

Assuming that the resource being competed for is water, the fractions of the shared water resource acquired by trees and grasses in a mixture can be predicted using the "electrical analog" model for water flow. Water uptake is a good proxy for nitrogen capture as well, since nitrogen is predominantly taken up in the soluble inorganic forms, which require water to be transported to the root. The key "'resistance" to water transport through the soil-plant atmosphere system is the leaf resistance (r_{leaf}), which is inversely proportional to the leaf area, L($r_{leaf} = L/r_{stomata}$, where $r_{stomata}$ is the resistance to flow through the pores in a unit area of leaf surface, and has a median value of about 2.5 s/cm in both trees and grasses in savannas). Because the grass leaves are "in parallel" with the tree leaves as a pathway of flow between the soil and atmosphere, the instantaneous fraction obtained by the grass is given by

$$F'_{wg} = \frac{r_{st}}{L_t} / \left(\frac{r_{sg}}{L_g} + \frac{r_{st}}{L_t} \right) \tag{9}$$

if we assume they are both exposed to the same soil water supply (φ_{soil}) and atmospheric demand (φ_{air}).

The fraction of the water resource obtained over the entire growing season is the integral of F_{wg}. There is currently no analytical way of predicting the temporal pattern of tree and grass leaf development ("phenology"), since it depends on the particulars of rainfall and temperature in a given year, in interaction with the soil and any residual water in it. Some generalisations allow a numerical approximation to be made to the time-integral of F_{wg}. Despite a range in mean annual rainfall in savannas between about 300 and 1200 mm/y, in most cases the rainfall is concentrated into a growing season of about 6 months duration. The trees leaf out as soon as the rains begin (or even before, if there is residual water in the soil, provided the air temperature is warm enough) and reach full leaf area expansion within about two weeks. Within the canopy area, the maximum tree leaf area index (one-sided leaf area per unit ground area) is fairly consistent in savannas at 3 m^2/m^2 (Scholes et al. 2002). The grasses, on the other hand, need to grow their leaf area by using current photosynthesis. Once the rains begin, and commencing with a leaf area of about 0.001 m^2/m^2 provided by root reserves, the leaf area grows by about 2% per day. Grass leaf senescence sets in once the accumulated days of water-stress exceed about a week. The water deficit is in turn partly determined by the combined tree and grass leaf area. Thus, as the season progresses, the grass leaf area grows exponentially, then peaks and declines. Tree leaf loss only commences after about 14 days of water stress. For the specific climate data of a typical savanna site at Skukuza, South Africa, an empirical representation of the relationship between stand basal area of trees (X) and the fraction of the water resource captured by trees when growing in the same soil volume as grasses is

$$F_{wt} = \frac{X}{2+X} \tag{10}$$

The grass productivity in subhabitat 2 can then be estimated as

$$P_{g2} = a[(R-c)(1-F_{w,t})] \tag{11}$$

Again, the key point is that the relationship has a convex shape in relation to tree basal area for all reasonable assumptions regarding tree and grass phenology.

In subhabitat 1, the area directly under the tree canopy, there is in addition to the competition for water, a reduction in water supply and solar radiation due to interception by the tree canopy. About 5% of the rain falling on the canopy is intercepted (de Villiers 1981). The fraction of the incoming solar radiation that penetrates through the tree canopy to reach the grass layer (F_{rg}), assuming a random orientation of the leaves in the canopy, is

$$F_{rg} = e^{-0.5Lt} \tag{12}$$

This has two effects. Firstly, it reduces the rate of photosynthesis, which is a function of intercepted radiation. Secondly, it reduces the rate of evaporation from the soil and vegetation under the canopy, thus prolonging the period for which photosynthesis occurs (Scholes 1988; Belsky et al. 1993). The degree to which the latter effect is expressed depends on what fraction of the energy budget for evaporation in the subcanopy area is provided by radiation, and what fraction by the movement of hot, dry air across the shaded area from the adjacent intercanopy areas. This will vary in a complex way with the relative water availability in the two patches. As a first approximation, I assume that the fraction provided by direct radiation is equal to A_1. Thus the "effective wetness" modifier of net rainfall (E) is proposed to be

$$E = \frac{1}{1 - A(1 - F_{rg})} \tag{13}$$

The relationship between radiation (I) and photosynthesis (P) can be approximated by a rectangular hyperbola

$$P = \frac{\alpha I P_{max}}{\alpha I + P_{max}} \tag{14}$$

Under the very bright conditions that apply in savanna regions, a large reduction in radiation translates into a smaller reduction in net primary production. The tree leaf area index in the area under the tree canopy is about 3, which reduces radiation by 70%, but production only by half. The shading reduction factor is given the symbol F_L.

The nutrient supply under the tree canopy is enhanced relative to the area between canopies. This is partly due to the fact that nutrients are taken up over the entire tree root extent, but preferentially deposited, by leaf fall and leaching, in the area below the canopy. From the perspective of grass production over the entire savanna plot, this is a "zero sum game" – the gains in the subcanopy habitat are losses elsewhere. The same applies to the additional nutrient input under tree canopies that derives from dust interception and the excreta of birds and animals congregating near the tree. However, a third component, due to nitrogen fixation by microbial symbionts exclusively associated with trees, is additional. If the fraction of the soil nitrogen in the soil under the tree that is derived from symbiotic nitrogen fixation is F_N (this value can be determined by inference from the proportions of isotopes of nitrogen present, and is seldom more than about 0.2), the productivity of grass beneath the tree canopy can be estimated as

$$P_{g1} = F_L(1 + F_n)aW(0.95R - c)(1 - F_{wt}) \tag{15}$$

The P_{g1} component has the potential to account for the observation that sometimes a small quantity of trees in the system leads to increased overall grass production. For high values of F_N and small values of stand basal area, the beneficial effects of nitrogen fixation can outweigh the detrimental effects of competition.

Note that the complementary relationships (the effect of stand basal area of trees on tree production) are generally concave. In other words, tree productivity rises steeply initially, and reaches an asymptote at high levels of X. A simple equation, consistent with the above logic, is that

$$P_t = (A_1 + A_2)F_{w,t}Ra_t \tag{16}$$

where a_t is the "rain use efficiency" of trees, about 0.35 g/m^2/mm.

3. Some Consequences of the Convexity

3.1. IS PRIMARY PRODUCTION IN TREE-GRASS SYSTEMS A ZERO-SUM GAME?

Historically, the managers of mixed tree-grass systems have tended to focus solely on only one of the components. If they were graziers, their objective was to maximise grass production, and if they were foresters, it was to maximise tree production. The typical solution, in the absence of a model that deals with the interaction of trees and grasses, is to eliminate the unwanted component. If the overall productivity of the savanna mixture was a linear function of the amount of tree in the system, then this still remains an "optimal solution": the system should be driven to whichever endpoint has the highest value. If the outcome of the interaction between trees and grasses is such that the highest productivity occurs at an intermediate level of tree cover, then it is possible that for some combinations of values associated with tree and grass products, a mixed system may be optimal.

The equations outlined above suggest that the only situation where the total production at intermediate tree density is higher than the productivity at either extreme is in the case where nitrogen fixation by trees occurs, and the rain-use efficiency of grasses is lower than that of trees (i.e., infertile soils).

There are cases where the landowner or society require *both* trees and grasses within the same landscape. One example is in ecotourism, where a lightly treed landscape ("parkland") is deemed more attractive than either a densely wooded one or a treeless one.

3.2. MULTIPLE STABLE CONFIGURATIONS

Historically, there have been two main schools of thought regarding the ecological dynamics of savannas (see review by Scholes and Archer 1996). At present, the schools are tending to converge on a hybrid concept. Both "traditional viewpoints", here contrasted for clarity, set out to explain the observed widespread coexistence of tree-grass mixtures. The "equilibrium school" posits that the competition between tree and grasses is balanced or limited – for instance by an even stronger competitive effect of trees on trees, or by an exclusive niche enjoyed by the otherwise weaker competitor. The "disturbance school" accepts that trees tend to out-compete grasses, but propose that fire (whose intensity is dependent on grass abundance) acts to keep the trees from completely dominating. The synthesis position argues that a comprehensive model of savanna dynamics requires both a competition mechanism and a disturbance mechanism.

Graphical analysis of the "state space" defined by the quantity of woody biomass (W) on one axis, and grass biomass (G) on the other axis, has been widely used to illustrate the existence of stable configurations (Walker, Ludwig, Holling and Peterman 1981; Walker and Noy Meir 1982). The isoclines where $\Delta G = 0$ and $\Delta W = 0$ (also called "nullclines") are plotted into this space, and their intersections with one another and with the axes are interpreted as stable or unstable. A recent analysis of resilience in savannas, based on a different model to the one proposed here but reaching similar conclusions, is given by Anderies et al. 2002.

The arguments of neither school are dependent on non-linearities, but the presence of non-linearities of the types described make certain outcomes more likely. For instance, following the "equilibrium school" it is theoretically possible to have a stable tree-grass mixture in a situation where there is a linear relation between grass production and tree amount. It is also possible to postulate a linear system with two stable end-points (complete domination by trees or treeless) separated by an unstable saddle-point. It is not possible, with purely linear forms, to have the mostly widely-observed situation, which is alternate configurations of the mixture in which both trees and grasses persist, though perhaps in small amounts (Figure 3). The non-linear effect of trees on grasses described by the equations in the first part of this paper allow the grass biomass to be severely reduced by tree biomass, but never completely eliminated from the mixture.

The production equations presented above make no prediction of an effect of grasses on trees. The empirical evidence on this topic is mixed: in an experiment where the grass cover was physically removed, tree growth was enhance in one vegetation type but not another (Knoop and Walker 1984). It is clear that the interaction is highly asymmetric: trees have a strong competitive effect on grasses, but grasses have a weak competitive effect on mature trees, although they may have a strong effect on saplings that have not grown above the grass layer.

The system of production equations presented above translates into the diagram shown in Figure 4, with suitable flux-to-stock conversions. The wood nullcline is a vertical straight line, since wood biomass is unaffected by grass biomass. This system has only one stable configuration: a tree-grass mixture, typically at a relatively high tree biomass and low grass biomass.

A quantitative analysis of the predictions of the disturbance school requires the introduction of an equation predicting the effect of grass biomass, via fire intensity, on tree biomass. A simple analysis is that fire intensity is directly proportional to the fuel load in the fire season, which is predominantly dry grass. Fire weather also influences intensity (Trollope and Potgieter 1985), but it is relatively consistent during the main burning season. Fire in savannas seldom kills trees (or grasses), but if the tree crown is in the flame zone, the aboveground parts are killed and must resprout from ground level. The critical threshold for recruitment of trees, and thus the long-term growth of the tree biomass, is whether the tree can grow beyond the flame height in the interval between fires. Flame height is an exponential function of fire intensity (van Wilgen 1986). For many semi-arid savannas, with a fire return interval of three to five years, it is empirically observed that a fire intensity in excess of 3000 W/m/s is required to suppress trees, which translates to a fuel load of around 300 g/m^2 under average burning conditions and a flame length of about 3 m. Introducing this simple relation into the graphical analysis results in Figure 5, in which two configurations are possible provided that the fuel accumulation over the mean inter-fire interval exceeds the 300 g/m^2 threshold. The first is the same high tree biomass-low grass biomass mixture predicted with competition alone, and the second has high grass biomass but no trees. The two are separated by an unstable "saddle point": trajectories to the left of this point tend to the treeless state (referred to as the "fire trap").

In practice, the "treeless" configuration usually has a scattering of mature trees that became established, often as a cohort, during a period when the system did not exceed the killing threshold. This can occur, for instance, during a prolonged period of below-average rainfall. In addition, there are typically many suppressed small trees within the grass layer, some of which may be decades old. The resilience

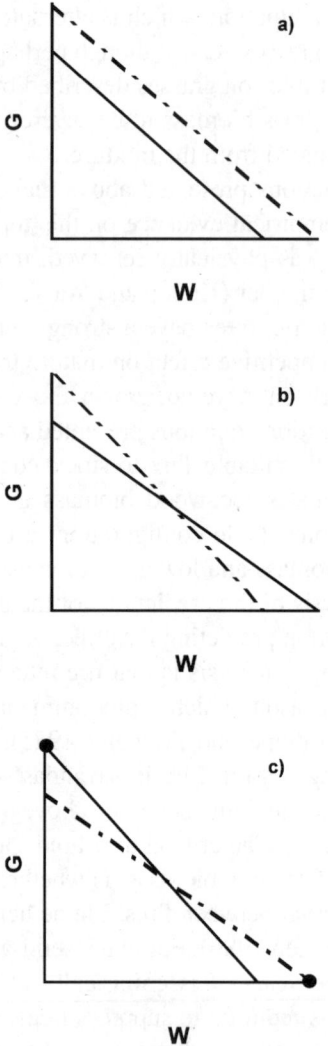

Figure 3. Graphical analysis of "nullclines" in the woody (W) versus grassy (G) state space is a widely used technique for analysing the stability states of savannas. The woody nullcline (dashed line) is the set of values of G at which $dW/dt = 0$, and the grass nullcline (solid line) the set of values of W where $dG/dt = 0$. The linear case is used here to illustrate the method. Where the tradeoffs between woody biomass and grass biomass are both linear, three qualitatively different cases are possible. In case (a) the tree nullcline lies above the grass nullcline over the entire range, and only one outcome is possible: total competitive exclusion of grasses by trees. (An analogous case is theoretically possible but not illustrated, with complete exclusion of trees by grasses.) In case (b) the woody nullcline lies above the grass nullcline at low wood biomass, but above it at high wood biomass. This is equivalent to saying that the competitive effect of trees on other trees is stronger than the effect of trees on grasses, and the same for grasses. The outcome is a stable coexistence of trees and grasses at an intermediate level. Case (c) has the grass nullcline above the wood nullcline at low W, and vice versa at high W. The outcome is an unstable ("saddle point") equilibrium at the intersection. If perturbed from this point the system will either trend to a treeless or grassless endpoints.

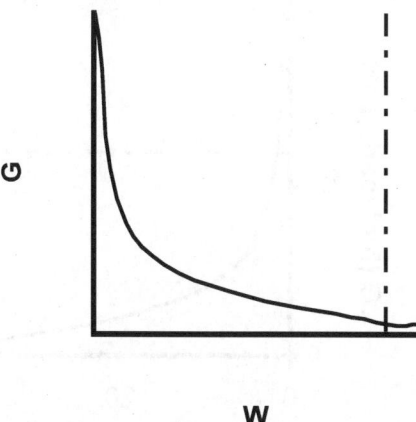

Figure 4. The state-space analysis that would result from the production equations described in this paper would show a strong, convex effect of woody biomass on the grass nullcline, but no effect of grass biomass on the wood nullcline. The wood nullcline would lie at the point where tree-on-tree competition limited further tree growth. The axes have not been quantified because the equations described apply to production at the annual timescale, not the instantaneous production and loss rates of the components. In reality, if the value of G is quantified as, for instance, aboveground green biomass, it shows a strong within-season variation from near-zero in the dry season, to a peak value in the middle of the wet season. The values of W, if expressed as total aboveground biomass of trees, vary slowly between years, but if expressed as tree leaf area (more meaningful for this purpose) show a strong, bounded variation within each year.

of this "open savanna" configuration can be measured as the distance between the treeless stable point and the unstable saddle-point. The effect of strong non-linearity in the effect of trees on grass production is to shift the saddle point to the left, reducing the resilience of the treeless configuration, and increasing the resilience of the high tree – low grass biomass configuration.

3.3. THE DYNAMICS OF BUSH ENCROACHMENT

When the tree-grass ecological system is coupled to a human system (pastoralism), it is commonly observed that the ratio of trees to grasses changes, often very rapidly. When European ranchers colonised the savannas of Africa, Australia, South and North America, the formerly open, grassy configuration frequently switched, within thirty to fifty years, to a densely woody configuration in which there was insufficient grass production, given the size of the holdings, to meet the income requirements of the rancher. Furthermore, the dense bush increases the cost of mustering, and in the extreme case, makes the residual grass physically inaccessible to the cattle. Interestingly, these problems were unknown in the African pastoral system which preceded the colonial period, and are rare on the communal grazing lands in Africa, despite the sometimes high livestock numbers they support. The communal lands often exhibit the opposite problem, "defore-

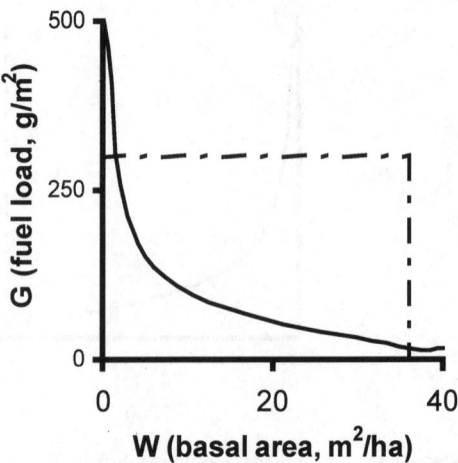

Figure 5. Addition of the effect of fire, largely fuelled by dead grass, on trees allows a state-space analysis based on production equations to be performed. It is assumed, in this case, that a fire of intensity 3000 kW/m/s (equivalent to flames 2.7 m in length, or a fuel load of 300 g/m^2) once every three years will prevent tree recruitment. This defines the horizontal portion of the wood nullcline. The vertical section is defined by the limit of woodiness in the absence of fire. An equation resulting from a meta-analysis of global tree cover data in savannas suggest this envelope to lie at $X_{max} = 1.2(R-300)$. The annual production equations can be used to generate the grass nulcline, by calculating the fuel accumulation over the fire interval (assuming that 60% of fuel carries over between years if not burned or eaten), as a function of woodiness, here represented by the tree basal area. The solution depicted here assumes 600 mm annual rainfall, 80% sand content in the soil and 20% of nitrogen being fixed by trees. At low W, sufficient grass accumulates in three years to exceed the tree mortality threshold. Therefore the system tends to an open state. At high W, insufficient grass fuel accumulates to control the trees, and the system tends to a closed state with a small amount of residual grass.

station", or a reduction in tree cover to a point where it threatens the supply of tree-based products.

There are several key differences between the two forms of management: ranchers generally suppressed fires, made little direct use of trees, and greatly reduced the numbers of browsers (herbivores that eat trees) in the system. In contrast, communal graziers deliberately set fires, harvest the trees for construction of huts, pens and for fuel, and typically run herds of goats in conjunction with cattle.

In terms of the simple graphical analysis, the effect of increased grazing is to lower the position of the grass nullcline, thus reducing the resilience of the "open" configuration, and eventually eliminating it as a possibility. The effect of long-term fire exclusion is to remove the horizontal segment of the wood nullcline (i.e., a return to Figure 4), which also eliminates the possibility of a stable "open" configuration. Increased fire frequency, up to a point and in the medium term, can lower the position of the horizontal portion of the wood nullcline, increasing the size of the fire trap. However, over-frequent fire could, in theory, run down the nutrients in the system or otherwise weaken the grass such that the grass nullcline is also lowered. Browsing and wood harvest have the effect of shifting the vertical portion of the wood nullcline to the left, meaning that the residual grass in the higher tree biomass configuration is increased, and the resilience of the woody configuration is reduced. Under very heavy browsing or wood harvesting the "mixture" configuration can be eliminated. These patterns are at least consistent with observations, even though they do not prove the hypothesis.

The ease and rapidity with which ranchers can tip the tree-grass configuration, and their difficulty in returning to the formerly more open state, is thus strongly related to the convexity of the effect of trees on grass. Following disturbance of the grass sward by grazers and fire suppression, small trees become established. The direct or indirect reduction of fire frequency and intensity and the exclusion of browsers allows them to grow, in a space of about a decade, to the point of being able to suppress grass production. The herd size is not reduced as the rancher begins to feel the economic pinch, resulting in even less grass being available to carry a fire, and the combined pastoralist-savanna system spirals into the encroached condition.

3.4. THE OPTIMAL PATTERN OF TREE CLEARING

A logical consequence of the convexity of the tree-grass relation is that if tree clearing is undertaken for the purpose of increasing grass production, and the amount of money available is insufficient to clear completely (or some trees are desired to be kept in the system for other reasons), the most cost-beneficial pattern of clearing is to remove all the trees in a portion of the landscape, rather than the remove a portion of the trees in all of the landscape. The guideline is to begin with the *least* encroached areas first (rather than the usual "intuitive" practice of tackling the most densely-treed areas first). It is generally easier to manage a "patchy clearing" pattern, since a clear operating rule can be established, fire can be used as a control mechanism, and the dispersal of tree propagules into the cleared area is reduced.

References

Anderies, M., M. A. Janssen and B. H. Walker (2002), 'Grazing Management, Resilience and the Dynamics of a Fire Driven Rangeland', *Ecosystems* **5**, 23–44.

Aucamp, A. J., J. E. Dankwerts, W. R. Teague and J. J. Venter (1983), 'The Role of *Acacia karroo* in the False Thorveld of the Eastern Cape', *Journal of the Grassland Society of Southern Africa* **8**, 151–154.

Beale, I. F. (1973), 'Tree Density Effects on Yields of Herbage and Tree Components in Southe-west Queensland Mulga (*Acacia aneura* F. Meull.) Scrub. *Tropical Grasslands* **7**, 135–142.

Belsky, A. J., S. M. Mwonga, R. G. Amundson, J. M. Duxbury and A. R. Ali (1993), 'Comparative Effects of Isolated Trees on Their Undercanopy Environments in High and Low Rainfall Savannas', *Journal of Applied Ecology* **30**, 143–155.

de Villiers, G. du T. (1981), 'Net Rainfall and Interception Losses in a *Burkea africana-Ochna pulchra* Tree Savanna', *Water SA* **7**, 4–25.

Donaldson, C. M. and D. M. Kelk (1970), 'An Investigation of the Veld Problems of the Molopo Area. 1. Early Findings', *Proceedings of the Grassland Society of Southern Africa* **1**, 27–32.

Groot, J. J. R. and A. Soumare (1995), Root Distribution of *Acacia seyal* and *Sclerocarya birrea* in Sahelian Rangelands', *Agroforestry Today* **7**, 9–11.

Knoop, W. T. and B. H. Walker (1984), 'Interactions of Woody and Herbaceous Vegetation in Two Savanna Communities at Nylsvley', *Journal of Ecology* **73**, 235–253.

le Houerou, H. N. (1984), 'Rain Use Efficiency: A Unifying Concept in Arid Land Ecology', *Journal of Arid Environments* **7**, 213–247.

Rutherford, M. C. (1980), 'Annual Plant Production-precipitation Relations in Arid and Semi-arid Regions', *South African Journal of Science* **76**, 53–56.

Scholes, R. J. (1988), *Response of Three Semi-arid Savannas on Contrasting Soils to Removal of the Woody Component*. PhD thesis, University of the Witwatersrand, Johannesburg.

Scholes, R. J. (1993), 'Nutrient Cycling in Semi-arid Grasslands and Savannas: Its Influence on Pattern, Productivity and Stability', *Proceedings of the XVII International Grassland Congress* (pp. 1331–1334). Palmerston North: International Grasslands Society.

Scholes, R. J. (1997), 'Savanna', in R. M. Cowling, D. M. Richardson and S. M. Pierce, eds., *Vegetation of southern Africa* (pp. 258–277). Cambridge: Cambridge University Press.

Scholes, R. J. and S. Archer (1997), 'Tree-grass Interactions in Savannas', *Annual Review of Ecology and Systematics* **28**, 517–544.

Scholes, R. J., P. G. H. Frost and Yuhong Tian (2002), 'Canopy Structure in Savannas along a Moisture Gradient on Kalahari Sands', *Global Change Biology* (in press).

Scholes, R. J. and D. O. Hall (1996), 'The Carbon Budget of Tropical Grasslands, Savannas and Woodland', in A. I. Breymeyer, D. O. Hall, J. M. Melillo and G. I. Agren, eds., *Global Change: Effects on Coniferous Forests and Grasslands* (pp. 69–100). SCOPE 1996. New York: John Wiley.

Scholes, R. J. and B. H. Walker (1993), *An African Savanna: Synthesis of the Nylsvley Study*. Cambridge: Cambridge University Press.

Trollope, W. S. W. and A. L. F. Potgieter (1985), 'Fire Behaviour in the Kruger National Park', *Proceedings of the Grassland Society of Southern Africa* **2**, 17–23.

van Noordwijk, M., G. Lawson, A. Soumare, J. J. R. Groot and K. Hairiah (1996), 'Root Distribution of Trees and Crops: Competition and/or Complementarity', in C. K. Ong and P. Huxley, eds., *Tree-Crop Interactions: A Physiological Approach* (pp. 319–364). Walingford: CAB International.

van Wilgen, B. W. (1986), 'A Simple Relationship for Estimating the Intensity of Fires in Natural Vegetation', *South African Journal of Botany* **52**, 384–385.

Walker, B. H., D. Ludwig, C. S. Holling and R. S. Peterman (1981), 'Stability of Semi-arid Savanna Grazing Systems', *Journal of Ecology* **69**, 473–498.

Walker, B. H. and I. Noy-Meir (1982), 'Aspects of the Stability and Resilience of Savanna Ecosystems', in B. J. Huntley and B. H. Walker, eds., *Ecology of Tropical Savannas* (pp. 556–609). Berlin: Springer.

Walker, J., R. M. Moore and J. A. Robertson (1972), 'Herbage Response to Tree and Shrub Thinning in *Eucalyptus populnea* Shrub Woodlands', *Australian Journal of Agricultural research* **23**, 405–410.

Walter, H. (1971), *Ecology of Tropical and Subtropical Vegetation*. Edinburgh: Oliver & Boyd.

Managing Systems with Non-convex Positive Feedback

W.A. BROCK and D. STARRETT
1986 S. 3rd East, P.O. Box 150821, Ely NV, 89315, U.S.A. (E-mail: Davstarret@aol.com)

Abstract. We study here optimal management of dynamic ecological systems that exhibit a destabilizing positive feedback. The prototype example is that of a shallow lake in which phosphorous loading placed by anthropogenic activities (fertilizers for farming and gardening) is stored in sediments until a critical level is reached after which there is a destabilizing return to the water – the tradeoff between farming interests and lake quality generates an optimal control problem. We show that in such systems, there may be a variety of local optima and associated basins of attraction wherein the optimal path may depend on starting state (phosphorous stock). We characterize the various possible optimal behaviors and identify the ambiguities that can only be resolved by choice of functional form.

Key words: basins of attraction, dynamic optimization, lake ecology, non-convex dynamics, positive feedback

1. Introduction

This article gives a fairly complete theoretical treatment of a deterministic version of the Carpenter et al. (1999) lake management model. This model is meant to be a general abstraction of optimal management problems for ecosystems where there is a phosphorous load placed by anthropogenic activities (fertilizers for farming and gardening) that contributes to stock stored in a lake. When that variable becomes too high, it sets off an internal positive feedback mechanism which impairs the ecosystem's ability to absorb and biodegrade loadings. Furthermore there are disamenities from degraded lake quality whose flow is proportional to the level of the stock variable. Here we will emphasize a fully optimizing version in which the manager can measure the stock and can control the loadings as function of stock. He chooses these loadings to best trade off the conflicting interests of farmers and lake users. Although we will use this lake model as the paradigm example, there are a number of other examples coming from ecological studies that seem to have similar structure and therefore ought to have roughly identical analysis. We will comment further on the general applicability of our analysis in the conclusions.

We show how methods of infinite horizon optimal control theory developed for problems with convex-concave dynamics of the state variable may be applied to the lake problem. A reasonably complete analysis turns out to be quite subtle.

Before we begin, we must say a few words about the style of exposition we shall use in this article. We shall be getting into problems of global analysis where there are many possibilities to examine and where the analysis will get bogged down quickly if we insist on writing in the completely rigorous "ε-δ" style. Hence, we shall write in the style of global analysis where arguments which can obviously be made rigorous will be exposited with geometrical devices like phase diagrams. Furthermore to avoid technical boundary conditions, we shall assume right hand and left hand "Inada" conditions and we shall assume existence of optima to infinite horizon control problems even though this has not been proved in all cases.

Existence theory is essentially trivial in discrete time given literature like that treated in Becker and Boyd (1997, see especially the references to Boyd's work on existence in continuous time problems). Existence theory is more technical in the continuous time case but is treated in Carlson, Haurie and Leizarowitz (1991).

2. The Stylized Lake Ecology

Following Carpenter et al. (1999), consider the following model of the dynamics of a lake where the limnology is boiled down to a one dimensional ordinary differential equation,

$$dx/dt = a - bx + f(x) = F(a, x), \quad x(0) = \text{given}. \tag{2.1}$$

Here, x represents the state of the system, and corresponds to the stock of phosphorous suspended in algae (which we will take as proportional to undesired turbidity of the water in the lake), a is phosphorous inputs from the watershed (sometimes referred to as the "loading"), b is the rate of loss per unit stock (sedimentation, outflow, and sequestration in organisms other than algae), and f(x) is internal loading. This function is assumed to be "S-shaped"; that is, for low stocks of phosphorous, additions tend to be stored in the lake bed so there is a relatively low marginal return to the water, whereas for higher stocks this marginal return increases, only to fall again when maximal suspension is approached. This equation represents a "minimal state variable" approximation to the complex foodweb of a real lake. The reduction and an argument that a convex-concave "recycling curve", f(x) is a good abstraction capturing real world positive feedback processes that are triggered inside a typical lake when P-load becomes too high is given by Carpenter et al. (1999) and the references in that paper. See Cottingham and Carpenter (1994, p. 2129) for a much more disaggregated model of the state variable dynamics which consists of a set of differential equations to represent the food chain of the lake. As we said before, one might think of (2.1) as an "aggregative reduced form" for the true disaggregated model.

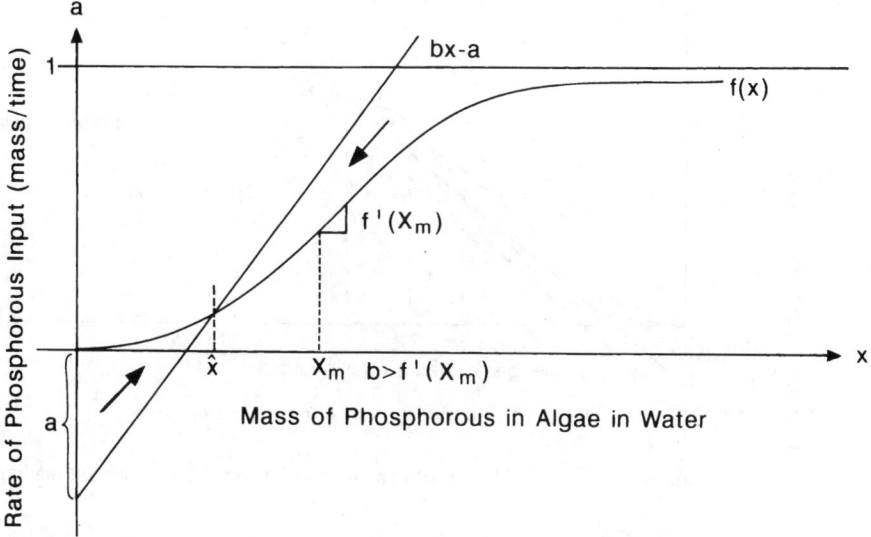

Figure 1. Lake model Case 1: unique equilibrium wiht constant loading.

3. Behavior under Constant Loading

We can get some intuition for the functioning of this model and set a benchmark for comparing with optimal management by considering behavior under constant loading, that is the dynamics of equation (2.1) when the loading rate from outside sources (a) is held constant. Depending on the relationship between the purification parameter (b) and the feedback function (f(x)), various qualitative behaviors are possible; we distinguish three distinct cases.

Case I: Unique equilibrium

When the purification rate is greater than the maximal feedback rate (b > max f'(x)) there will be a unique long run equilibrium steady state for any given constant loading rate and the steady state will be a continuous increasing function of the loading. Furthermore, convergence to the steady state will be monotone from any starting stock of phosphorous. Referring to Figure 1, if the loading is a and x is above $\hat{x}$, bx-a is larger than f(x) so dx/dt is negative and x will fall toward the unique steady state $\hat{x}$. Similarly, the reverse dynamics holds if initial stock is below $\hat{x}$. Here the nonconvexity appears inessential in that behavior is qualitatively as it would be in a fully convex version.

However, we will see that this need not be true in the optimizing version – that is for some combinations of parameters consistent with Case I, there will be multiple local optima. As we will see, this situation becomes more likely in the other cases.

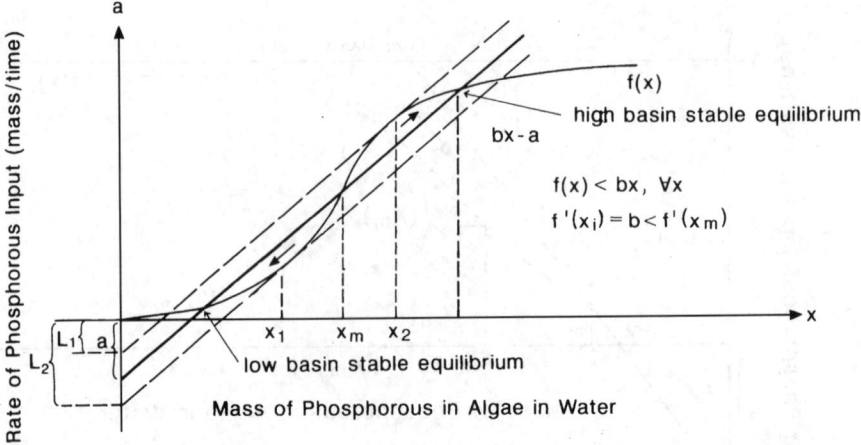

Figure 2. Lake model Case 2: 2 basins of attraction with reversibility (constant loading).

Case II: Multiple equilibria with reversibility

When the purification rate is smaller than the maximal feedback rate, there will be multiple long run steady states for some values of the constant loading. Case II corresponds to the situation when, in addition, $f(x)$ is less than bx for all x (see Figure 2). Then for initial loadings between L_1 and L_2, there will be two locally stable long run steady states, one involving a relatively low x "clear water" state (referred to as oligotrophic in the ecology literature) which is highly valued by the lake's users, but also a relatively high x "turbid water" state (referred to as eutrophic in the ecology literature) which is disliked by the lake's users but valued by agricultural interests because it allows them high fertilizer use. Which state is reached for a given constant loading depends on the initial stock of phosphorous, high initial stocks leading to a eutrophic long run equilibrium and vise versa (as indicated by arrows in Figure 2). These two "basins of attraction" of the lake will play a major role in the following discussion. In this situation, history matters in that for a given constant load the long run steady state will depend on the initial level of x. However, all long run outcomes are reversible in this case. A sufficiently low load (below L_1) will always restore the clear water type state if it is maintained sufficiently long.

We will show that a fully optimizing version of case II can display a vide variety of qualitative behaviors. There can be a single locally optimal path or as many as you like, depending on parameters. Further, we will show that history may or may not matter in the optimizing solution.

Case III: Multiple equilibria with irreversibility

If $f(x)$ lies above bx for sufficiently high x, then behavior is as in case II except that once the phosphorous level exceeds a certain critical value, there is no policy

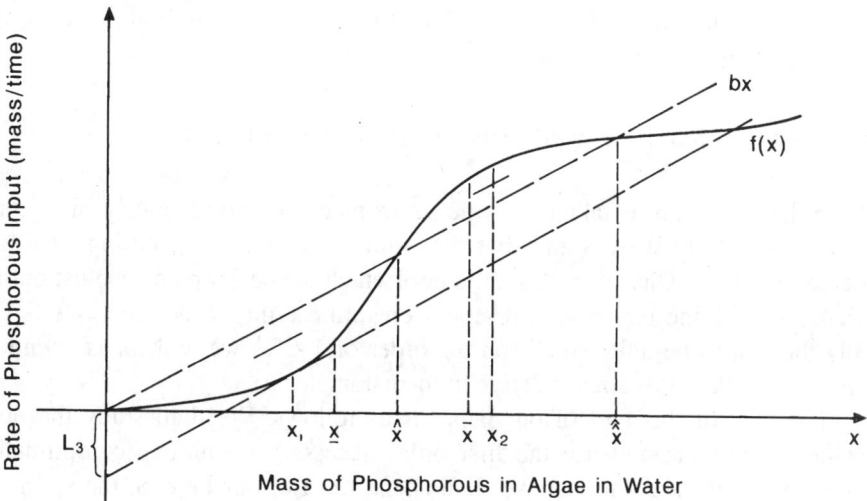

Figure 3. Lake model Case 3: multiple equilibria with irreversibility.

that can ever again achieve lower values. In Figure 3, this critical level is $\hat{x}$. If the initial stock is larger than that, then even if we reduce the external loading to zero, the phosphorous stock will converge to a eutrophic long run steady state (at $\hat{x}$).

In our optimizing version there will be a unique locally optimal policy if the initial stock is above $\hat{x}$. However, if it starts below, there may still be multiple local optima – that is it might be desirable to cross the threshold from below in an optimal management scheme. We will try to show when and how this happens.

Notice that the function f(x) is the source of the potential hysteresis in all of these cases. A plausible mechanism to generate an effect like f(x) is the theory put forth in Scheffer (1997, Chapter 3 and Figures 3.19 and 3.22). Further, the model's "prediction" that lakes may be seen to flip from low turbidity to high turbidity states in response to small changes in loading factors seems to be confirmed by observation.

4. An Optimizing Framework

There is a community of individuals who share the lake and its watershed. These individuals have conflicting interests. Some are producers whose activities release loadings into the catch basin of the lake, which contribute to the term a in (2.1). Others are users of the services of the lake and are injured by the "dumping" activities of the others in that they care about the overall quality of the lake which we proxy by the stock x – higher stocks meaning lower quality.

Utility (measured, for example, as increases in profitability of economic operations) is generated for the producers of a from agriculture, lake side cottages, developments, etc. Disutility is generated by x and must be measured by willingness to pay for a clean water environment.

We formulate the objective of a Social Planner who acts in the interest of the community as a whole as follows,

$$W(x) = \text{Max}\{\int_0^\infty \exp(-rt)\{U(a, x)\}dt\}, \text{ subject to (2.1)}. \tag{4.1}$$

Here the maximum is taken over the set of piecewise continuously differentiable functions $\{a(t)\}$. We assume that the utility $U(a, x)$, is increasing in a and is decreasing in x. This generates a tradeoff which the social planner must optimize. Here $r > 0$ is the real rate of discount on future utility. Since this rate is generally thought to be quite small (on the order of 1–2%) we will focus attention on situations where it is small relative to the parameter b.

The plan of this part of our paper is as follows. We shall study the optimal solutions of (4.1) and use the first order necessary conditions of optimality to characterize these solutions. We shall spend most of our time on the special case $U(a, x) = U_1(a) + U_2(x)$ because, here, we can "separate" the interests of the Affectors (those interests like agriculture and developers whose activities damage the resource) from the interests of the Enjoyers (those interests like fishers, viewers, and swimmers whose activities do not injure the resource). We take the terms "Affectors" and "Enjoyers" from the work of Scheffer et al. (1998). We shall also spend some time on the case where U_1 is linear because it generates a particularly simple dynamics that motivates the more general behavior. Linearity can be thought of as representing a reduced form restricted profit function of a constant returns to scale Affector sector. This case is mostly of pedagogical interest since with a fixed amount of land in proximity to the lake, we would expect diminishing returns.

5. The Most Rapid Approach Special Case

Here we study the linear, separable case $U_1 = Aa$, $U_2 = -c(x)$, $c' > 0$, $c'' > 0$. For the deterministic case linear U generates a Most Rapid Approach Problem (MRAP) where one operates the control variable a to push the state variable x to the maximum of a certain function of x as we shall see. Thus we now study the problem:

$$W(x) = \text{Max}\{\int_0^\infty \exp(-rt)\{Aa - c(x)\}dt\}, \text{ subject to} \tag{5.3}$$

$$dx/dt = a - bx + f(x). \tag{5.4}$$

We assume that $f(x)$ is convex-concave and is increasing in x with $f(0) = 0$, $f'(0) = 0$, $f'(\infty) = 0$, with unique $x_m = \text{argmax } f'$, $f'' > 0$ for $x < x_m$ $f'' < 0$ for $x > x_m$. Notice that if we define $AP(x) = f(x)/x$, $MP(x) = f'(x)$, then just as in elementary economics of increasing returns production functions, we have MP > AP (MP < AP) for $x < x_m$ ($x > x_m$) with MP = AP for $x = x_m$. Furthermore, $AP(0) = MP(0)$ and $AP(\infty) = MP(\infty) = 0$.

Now, for any finite T, substitute for "a" from (5.4) and observe that

$$\int_0^T \exp(-rt)\{Aa - c(x)\}dt = \tag{5.5}$$

$$A[\exp(-rT)x(T) - x(0)] + \int_0^T \exp(-rt)\{A[rx + bx - f(x)] - c(x)\}dt.$$

Hence if we restrict ourselves to the set of functions x(.) that satisfy the transversality condition:

$$\exp(-rT)x(T) \to 0, T \to \infty, \tag{5.6}$$

then (5.5) suggests that we control a(.) to move x(t) to most rapidly approach

$$x^* = \operatorname{argmax} \pi(x) = \{A[rx + bx - f(x)] - c(x)\} \tag{5.7}$$

Indeed, since the objective is now linear in control (a) we know from the work of Spence and Starrett (1975) that even if there are restrictions on the control, the optimum from any starting point is to most rapidly approach some local maximum in (5.7). We can think of this objective as trading off long run interests of Affectors and Enjoyers. Affectors get a flow benefit of rA per unit from the stock per unit time while Enjoyers pay cost c(x). In addition the stock provides a net purification benefit of bx − f(x) which acts like a productivity term in the objective.

Notice that it is easy for the mathematics in (5.7) to yield "unpleasant" solutions. For example, if the turbidity disutility $U_2(x) = -c(x)$ were bounded from below, (5.7) would shoot x off to infinity and turn the lake into a dumping ground for the a > 0 polluting "stakeholders". This is so because the polluters are willing to pay A > 0 for each extra unit of x-dumping while the willingness to pay to prevent such dumping embodied in the disutility is falling to zero.

Part of the reason for this anomaly derives from our assumption of linearity (marginal benefit from pollution assumed not to decline in pollution level). However, it seems sensible to posit an unbounded cost which will assure an interior solution. If x(t) can be moved immediately to x*, we have from (5.5)

$$W(x) = -Ax(0) + (1/r)[A[rx^* + bx^* - f(x^*)] - c(x^*)]. \tag{5.8}$$

An initially pristine lake is one with x(0) = 0. We may differentiate W to obtain how much an innovation would be worth that increased "b" by one unit,

$$dW/db = (1/r)ax^*. \tag{5.9}$$

The innovation allows the lake to "biodegrade" one more unit of x per unit time. This extra loading capacity is worth A to the polluters each period. When this amount is capitalized over all periods at rate r we obtain (5.9). Similar sensitivity analysis can be done with for other parameters of interest.

The analysis provides more insight into the nonlinear case if we consider constraints on the control variable that will prevent instantaneous adjustment.

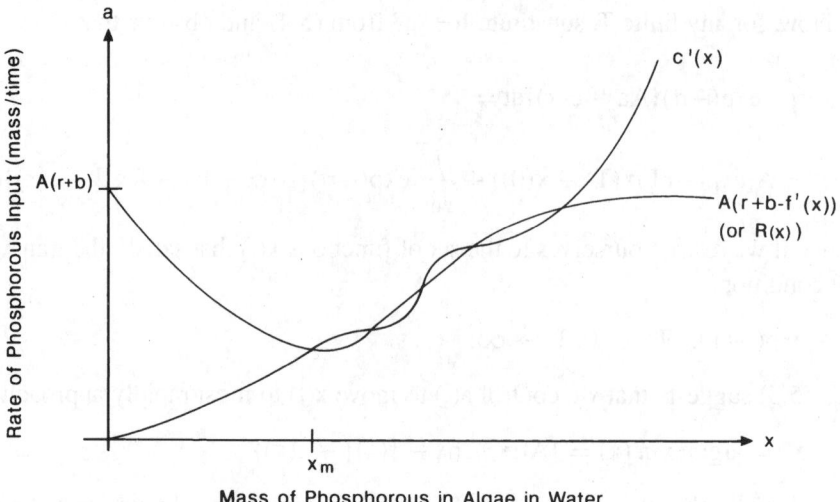

Figure 4. A multitude of equilibria.

Nonnegative values of a naturally prevent such downward adjustment and there may well also be upper bounds (say a < M) which similarly prevent upward adjustment. Then, as in Spence and Starrett (1975) we must search among local optima of $\pi(x)$. Examining this function, we see that there may well be multiple local optima. To see this, examine the first order conditions for a maximum:

$$A[r + b - f'(x)] = c'(x). \tag{5.10}$$

Since for $x > x_m$, $f'(.)$ is decreasing and $c'(x)$ is increasing, we can make particular choices for these functions that make the left and right sides of (5.10) equal as often as we like (see Figure 4).[1] (We will be somewhat more rigorous in demonstrating similar properties in the nonlinear case.)

As always, generically, the equalities will alternate between local maxima and local minima. Since we assume $f'(0) = c'(0) = 0$, it is easy to see that the first intersection must correspond to a local maximum; also, since $f'(\infty) = 0$ and $c'(\infty) = \infty$, the last intersection is also a local maximum. Of course, multiplicity need not always happen for all choices of parameters and functions, but we see that arbitrarily large numbers of local optima cannot be ruled out a priori. This observation will carry over to the general optimization problem as we will see.

Let us consider the simplest generic multiplicity case where $\pi(x)$ has two local maxima α and γ separated by a local minimum β. The remaining problem is to determine which of the local maxima is the optimal long run steady state from a given starting position given the restrictions on controls. The global maximum may not always be the long run optimum, as it may take too long to reach it from the other local optimum and this matters if we discount enough.

We will not try to solve this problem completely here. However, the basic qualitative analysis is easy to give and will carry over to our nonlinear general case. As long as the controls are bounded so that our single state variable must move continuously, then a simple application of the Bellman principle implies that it must move monotonically in an optimal trajectory and the set of states for which a particular local maximum is the long run optimum must be connected. (The set associated with a particular local maximum will be referred to as its basin of attraction.) Consequently, either the global maximum is always optimal or there is a knife-edge point between the local optima such that the global optimum goes to γ for all starting stocks to the right of the knife-edge and to α for all stocks to the left. We refer to this knife-edge point as the Skiba point since he identifies and studied it in a related renewable resource problem (see Skiba (1978)).

We turn now to the general nonlinear case where we will find a similar qualitative picture. We will try to discover what particular features of the problem lead to multiplicities (or lack thereof) and will discuss ways of determining Skiba points that result.

6. General Analysis of the Lake Problem

Consider the general separable-utility problem:

$$W(x) = \text{Max}\{\int_0^\infty \exp(-rt)\{u(a) - Bc(x)\}dt\}, \tag{6.1}$$

Subject to $dx/dt = a - bx + f(x)$, $x(0) = x$.

We assume: $u' > 0$, $u'(0) = +\infty$, $u'' \leq 0$, $c(0) = 0$, $c' > 0$, $c'(0) = 0$, $c'' \geq 0$. Verbally, utility to Affectors is concave (diminishing marginal utility), costs to Enjoyers are convex (increasing marginal cost), with B representing a parameter that determines the relative importance of the two groups. The assumptions of $f(.)$ are as in the previous section.

We use optimal control theory to characterize optimal solutions. Using p as the current value costate variable, we set up the current value Hamiltonian:

$$H(a, x, p) = u(a) - Bc(x) + p\{a - bx + f(x)\}.$$

Then the maximum principle tells us that optimal loading is set to satisfy

$$a > 0 \rightarrow u'(a) = -p, \quad a = 0 \rightarrow u'(a) < -p, \tag{6.2}$$

defining a function a(p). The Hamiltonian dynamics then are written as the coupled pair of differential equations:

$$dx/dt = a(p) - bx + f(x), \tag{6.3}$$

$$dp/dt = \{r + b - f'(x)\}p + Bc'(x). \tag{6.4}$$

The equations (6.2)–(6.4) constitute necessary conditions for a solution to (6.1) in that for the optimal policy we must be able to find a function p(.) such that these equations are satisfied for all time.

Parenthetically we remark that the assumption, $u'(0) = +\infty$ implies that $a = 0 \to -p = +\infty$, so we may ignore the strict inequality case in (6.2). We shall do this in what follows. Also, due to this fact, it is sometimes convenient (particularly in doing phase diagrams) to change variables and work in $\{a, x\}$ space rather than $\{p, x\}$ space. This is accomplished by logarithmically differentiating (6.2) with respect to time, obtaining:

$$(dp/dt)/p = (da/dt)[u''(a)/u'(a)]. \tag{6.5}$$

Then, substituting into (6.4). We get the corresponding pair of equations:

$$dx/dt = a - bx + f(x) \equiv X(a, x), \tag{6.3'}$$

$$da/dt = [r + b - f'(x)][u'(a)/u''(a)] - B[c'(x)/u''(a)] \equiv A(a, x, B). \tag{6.4'}$$

Note for future reference that $\partial A/\partial B = -c'(x)/u''(a) > 0$.

We shall proceed to do a general analysis of this problem by carrying out the following steps: (i) Investigate sufficient conditions for a unique steady state in these equations and find sufficient conditions for multiple steady states. It turns out to be easy to construct examples where there are as many steady states as you want. (ii) Conduct a dynamic analysis using linearization around steady states, $p(x) = W'(x)$, and phase diagrams in order to piece together the value function $W(x)$ following the strategy of Skiba (1978), Brock and Dechert (1983, 1985) (exposited in Brock and Malliaris (1989)).

7. A Unique Steady State

It turns out that there can be a unique locally optimal steady state in either case one or case two as identified earlier (but not in case three). However the outcome is quite parameter dependent and with particular choices, there can be multiple steady states in any of the cases. In this section, we restrict ourselves to cases one and two. We show first that for any parametrization consistent with these cases, there is a breakeven B (call it B*) such that there exists a locally optimal steady state in the eutrophic (high x) basin of attraction if an only if B is below B*. It will then follow almost immediately that if B is above B*, there is a unique local (and therefore global) optimal path converging to a steady state in the oligotrophic (low x) basin of attraction. We will also see that this path is saddle-value stable. In later sections, we will explore the behavior with multiple steady states.

To identify steady states, we look for a joint solution to the equations $dx/dt = da/dt = 0$, namely:

$$a - bx + f(x) = 0, \tag{7.3'}$$

$$[r + b - f'(x)][u'(a)] - B[c'(x)] = 0. \tag{7.4'}$$

These equations define curves in a,x space where $dx/dt = 0$ and $da/dt = 0$ respectively. Note that in this situation, both of these define a uniquely as a function of x and we label these $a_1(x)$ and $a_2(x)$ for convenience. Since $bx > f(x)$, all $x > 0$ in cases one and two, $a_1(x)$ will be strictly greater than zero for all positive x in these cases. This is the important contrast with case three for purposes of this section. Any candidate for an optimal steady state must satisfy (7.3'), (7.4'), since (6.3'), (6.4') are Hamiltonian necessary conditions.

Recall the following features of f(.): x_m is the unique inflection point where $f'' = 0$. Where they exist (case two), x_1 and x_2 are the two interior points where $f'(x_i) = b$.

Lemma 1:
In case two, there exists a B such that for all larger B, and any (a, x) with $x \geq x_1$ and $X(a, x) = 0$, then $A(a, x, B) > 0$.

Proof:
Examining (6.3'), (6.4'), we can see that the conditions $X(a, x) = 0$, $A(a, x, B) > 0$ are equivalent to

$$[r + b - f'(x)][u'(bx - f(x))] - B[c'(x)] < 0. \tag{7.5}$$

Now, for $x \geq x_1$, using concavity of u(.), convexity of c(.) and nonnegativity of f(.),

$$[r + b - f'(x)][u'(bx - f(x))] - B[c'(x)]$$
$$< [r + b][u'(bx_2 - f(x_2))] - B[c'(x_1)].$$

Hence, recalling that $\partial A/\partial B > 0$, the conditions of the lemma are satisfied by choosing

$$B = [r + b][u'(bx_2 - f(x_2))]/c'(x_1). \quad \text{Q.E.D.}$$

Corollary 1:
In case one, there exists a B such that for all larger B, and any (a, x) with $x \geq x_m$ and $X(a, x) = 0$, then $A(a, x, B) > 0$.

Proof:
Observe that all the steps of Lemma 1 go through for case one if x_m is substituted for both x_1 and x_2. Q.E.D.

The significance of lemma and corollary for us is that when B is large enough there cannot be any candidate steady states with x bigger that x_1 (case 2) or x_m (case 1). This accords with our intuition that if enjoyers matter enough the social planner will not want to go toward the "high x" basin of attraction. Next we establish something of the reverse for low B.

Lemma 2:
Pick any $\hat{x}$ bigger than x_m for which $r + b > f'(\hat{x})$.[2] Then in either case one or case two, if $\hat{a}$ is such that $X(\hat{a}, \hat{x}) = 0$, there exists $\underline{B} > 0$ with $A(\hat{a}, \hat{x}, \underline{B}) < 0$.

Proof:
We require

$$[r + b - f'(\hat{x})][u'(b\hat{x} - f(\hat{x}))] > \underline{B}[c'(\hat{x})]. \tag{7.5}$$

Since the left side of (7.5) is strictly positive and $c'(\hat{x})$ is finite, this inequality can be satisfied for positive $\underline{B}$. Q.E.D.

Using these lemmas, we can show that there is a breakeven relative importance of the parties such that if the enjoyers are relatively more important there will be a candidate steady state in the low-x basin of attraction whereas if they are less important there will always be a candidate steady state in the high-x basin of attraction.

Proposition 1: (note: This and later propositions to be read twice – with and without braces)
In case 1 {case 2}, there exists a $B^* > 0$, such that

$$B \leq B^* \text{ if and only if there is a candidate steady state with } x > x_m\{x_1\} \tag{7.6}$$

Proof:
We set B^* to be the smallest that will satisfy lemma 1 {corollary 1}. Since $\underline{B}$ sets a positive lower bound, there must be such a positive least upper bound. Q.E.D.

Proposition 2:
In case 1 {case 2}, if $B > B^*$, there will exist a unique candidate steady state with $x < x_m$ {$x < x_1$}.

Proof:
From Corollary 1 {lemma 1} we know that

$$A(a_1(x_m), x_m, B) > 0 \ \{A(a_1(x_1), x_1, B) > 0\}$$

Further, since $a_1(0) = 0$, using (6.4') and the assumption $c'(0) = 0$, we see that

$$A(a_1(0), 0, B) < 0.$$

Hence by continuity, there exists $0 < x^* < x_m$ {x_1} such that $A(a_1(x^*), x^*, B) = 0$. By definitions of functions $a_1(.)$ and $A(.)$, x^* satisfies the steady state equations (7.3'), (7.4'). Further, since $a_1(.)$ is monotone increasing and $a_2(.)$ is monotone decreasing in x on the interval $[0, x_m]$ {$[0, x_1]$} there can be only one such candidate steady state in that interval. And from proposition 1, there are no such candidates outside the interval. Q.E.D.

In the remainder of this section we focus on the situation in which there is a unique steady state ($B > B^*$). The interpretation of this steady state in terms of long run optimization is very similar to what we found in the MRAP case. (7.4') tells us that the marginal consumption plus investment benefit of the stock to Affectors ($[r + b - f'(x)]u'(a)$) must be set equal to the marginal cost to Enjoyers ($B[c'(x)]$). The only difference now is that marginal utility to Affectors is no longer constant and in the long run optimum, a must satisfy (7.3′) in order that the steady state level of x be maintained.

Notice that for the "Ramsey" case, $r = 0$, steady states, (a^*, x^*) solve the problem:

$$(a^*, x^*) = \text{argmax } \{U(a, x), \text{ s.t. } 0 = a - bx + f(x)\}.$$

Substitute out $a = f(x) - bx$ from the constraint into the utility to obtain

$$x^* = \text{argmax } \{U(f(x) - bx, x)\}.$$

In our case of separable utility this objective becomes $u(f(x) - bx) - Bc(x)$ and looks even more like a benefit minus cost objective.

Now however, we care not just about the steady state level of x but also on how quickly we get there: The value per unit of a depends on the rate of loading so the total benefit to Affectors depends on how loads are distributed over time. So we turn next to transitions.

The behavior away from steady states is most easily (though not entirely rigorously) analyzed using phase diagrams. These are constructed by plotting the functions $a_1(.)$ and $a_2(.)$ And deducing the direction of motion in all regions of the a, x space between these curves using (6.3′) and (6.4′). In the following discussion, refer to Figures 5 (for case 1) and 6 (for case 2). Shapes of the curves differ in the two cases.

From (7.3′) we deduce:

$$da_1/dx = b - f'(x). \tag{7.7}$$

Thus, $a_1(.)$ is monotone increasing for case one, whereas it increases to x_1, decreases from there to x_2, and increases thereafter in case 2 (as we have drawn).

Turning to (7.4′), we see that (generally) in case 2, $a_2(.)$ will intersect the x axis. This happens as long as f′ is bigger than $r + b$ on some interval (represented in the diagrams as $[\underline{x}, \bar{x}]$). Implicitly differentiating to find shapes elsewhere, we find

$$da_2/dx = [Bc''(x) + u'(a)f''(x)]/[(r + b - f'(x))u''(a)]. \tag{7.8}$$

Outside the above mentioned interval the denominator of (7.8) is always negative. The numerator is positive for $x < x_m$, and also for sufficiently large x (since f′ goes to zero). However in an intermediate range, it is clearly ambiguous. Indeed, we know that for B sufficiently close to B^*, $a_2(.)$ must have a rising segment since at B^* the two curves must intersect and $a_1(.)$ is strictly positive (and strictly increasing in case 1). Since, in addition, computational simulations generally exhibit an increasing segment, we have drawn the diagrams in that way. This feature will have interesting implications as we will see below.

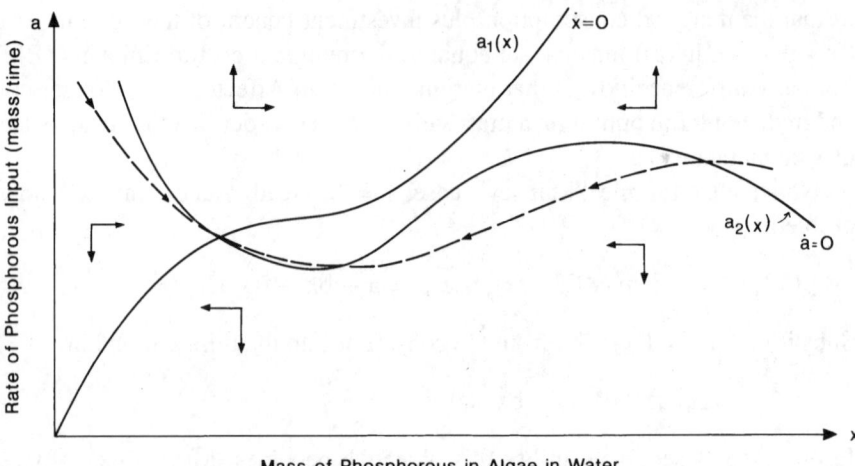

Figure 5. Phase diagram (B > B*, Case 1, unique equilibrium).

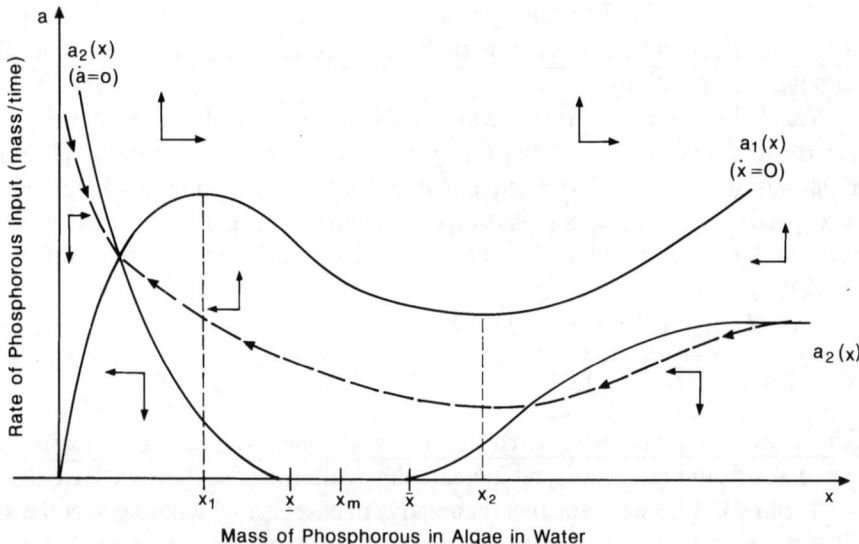

Figure 6. Phase Diagram: (B > B*, Case 2, unique equilibrium).

We have drawn in arrows to indicate the direction of motion off the curves of stationarity as determined by equations (6.3′) and (6.4′). For example, since $\partial A/\partial a > 0$. We know that dx/dt is positive above $a_1(.)$ and negative below. Similarly we find that da/dt is positive above $a_2(.)$ and negative below. Thus we get the indicated qualitative dynamics in the four "quadrants" of our diagrams.

Examining these diagrams, we see that the unique steady state is saddle value stable (this can be demonstrated more rigorously using linearizations, and we will utilize that approach later). Under standard regularity conditions (cf. Hartman (1973, See Chapter IX on Invariant Manifolds and Linearizations)), there will exist a unique one dimensional local stable manifold of points satisfying these differential equations and converging to the steady state. Hence by running time forward and backwards along solutions of the costate/state equations that lie on the stable manifold near the steady state, one may construct an entire global stable manifold. Since the costate variable p(x) must be the derivative of the value function, points (x, p(x)) on the global stable manifold give (x, V'(x)) for a "candidate" value function V(x).

Are there other solutions (besides that lying on the global stable manifold of the unique steady state depicted in Figures 5 and 6), of the costate/state equations that might be optimal? This question has not been definitively answered even in the general literature on optimal control. All we know in general is that if there is an optimum it must satisfy the costate/state equations for all x. However, arguments parallel to those in Skiba (1978) and Ekelund and Scheinkman (1986) suggest that we look for solutions of the costate/state equations that satisfy the transversality condition at infinity (TVC∞),

$$(\text{TVC}\infty) \ \exp(-rt)p(t)x(t) \to 0, t \to \infty. \tag{7.9}$$

Solutions of the costate/state equations that lie on the global stable manifold satisfy (7.9) as long as r > 0. This is because p and x converge to finite limits and the negative exponential converges to zero. However, since transversality is not known to be strictly necessary or sufficient here, one generally finds an auxilliary argument to rule out other paths.

If we assume c(x) increases sufficiently rapidly relative to u(a) this will eliminate (x(t), p(t)) such that x(t) → ∞, p(t) → 0, a(t) → ∞ by a comparison argument. Solutions (x(t), p(t)) such that x(t) → 0, a(t) → 0, p(t) → −∞ as t → ∞ can also be eliminated by a comparison argument provided, c(0) = 0 and u(0) < 0. Hence the only candidate solution remaining in this case is the one depicted in the figures. Consequently there is a one dimensional stable manifold which extends over all values of x, converging to the steady state and these paths constitute the optimal transition programs. As mentioned before (and now seen in the diagrams) the steady states must lie in the low-x basin of attraction and paths converge monotonically in x. What is more surprising and new is that they need not converge monotonically in a. One might expect that if the initial stock is too high relative to the long run optimum, the manager would initially set the loading quite low (thus lowering the stock quickly to benefit Enjoyers who experience increasing marginal cost) and let it rise over time to the long run optimum.

However, this behavior is not generally optimal here, and the reason has to do with the shape of f(.). When x is quite large, there is saturation and lowering x does not give the feedback benefits it will in intermediate ranges. Thus, it may be optimal to start with relatively high levels of a (possibly even higher than the long run optimum) to benefit Affectors, and lowering these only when the region of maximum feedback benefit is reached.

It is interesting to note that parameters r and B affect only the da/dt = 0 curve which makes for easy comparative dynamics. For example, an increase in the discount rate will raise the $a_2(.)$ curve but not affect the $a_1(.)$ curve. Consequently the long run equilibrium level of x will rise – that is, with more impatience, the planner worries more about the immediate benefits to Affectors and less to the long run benefits of Enjoyers.

Similarly, we can deduce comparative dynamics by considering (for example) a surprise increase in B due to a new surge of wealthy recreationists locating on the shores of the lake. This event causes the da/dt = 0 curve to fall and does not change the dx/dt = 0 curve. Let the lake be at the old steady state. Draw the old stable manifold on Figure 5 (or 6) and draw the new stable manifold which lies below it. The new optimal loading a drops immediately to locate on the new stable manifold and slowly *rises* as x falls to the new, lower steady state level of x. We italicize "rises" because one might expect "a" to keep decreasing for a while. In the real world there would be adjustment costs to rapid change in "a" which may deliver a result more consistent with intuition.

8. Multiple Steady States

Once we pass from the situation of a unique steady state, there generally will be more than one locally stable manifold satisfying all necessary conditions and transversality. Our strategy will be to use phase diagrams of the current value costate/state equations to identify all solutions that satisfy the transversality condition at infinity. Given these, we can use the Hamilton-Jacobi property: $p(x) = W'(x)$ to construct a finite collection of "candidate" value functions $\{V_i(x), i = 1, 2, \ldots, I\}$. The optimal value function is given by

$$W(x) = \max_i \{V_i(x)\}. \tag{8.1}$$

The points where the maximizing "i" changes constitute the switch (Skiba) points that correspond to those we discussed briefly in the MRAP section. We will try to characterize these later. Generally, each of the "i"s above will correspond to a locally stable steady state, so first we explore the potential multiplicity of these.

We return to an analysis of the steady state equation:

$$[r + b - f'(x)]u'(bx - f(x)) = Bc'(x). \tag{8.2}$$

If we let R(x) represent the function on the left hand side of (8.2), then the argument given in section 5 can be used to show that if (and only if) R(.) has a monotone increasing segment in an interval where it is strictly positive, a $c'(x)$ can be found under which there are arbitrarily many solutions to this equation. We demonstrate this somewhat more rigorously here.

Lemma 3:
Assume $R'(.)$ is continuous and there is an $x = A > 0$ such that $R(A) > 0$ and

$R'(A) > 0$. Then there is a convex increasing cost function $c(x)$ such that (8.2) has countably many positive solutions. Under the plausible right hand and left hand "Inada" conditions $R(0) > c'(0)$ and there is x_M such that $x > x_M \to R(x) < c'(x)$, we have (generically) R intersecting c' from above for the first steady state and R intersecting c' from above for the last steady state. Hence, generically there are an odd number of steady states if the number is finite.

Proof:
Notice that the only requirement for convexity of $c(x)$ is that $c'(x)$ be nondecreasing. Now $R'(A) > 0$ implies there is an open interval $I = (A - e, A + e)$, $e > 0$ such that $R(x)$ is strictly increasing on I. Now construct a nonnegative non decreasing function $c'(x)$ on $[0, \infty)$ such that it intersects $R(x)$ on I a countable number of times. This can be done because $R(x)$ is strictly increasing on I. The second part of the Lemma is obvious. Q.E.D.

Of course, assumptions on $R(.)$ are not fundamental since they are not stated on the primitives of the problem. We now seek such primitive assumptions. Note first that $R(.)$ is not well defined unless $bx > f(x)$ so we restrict to cases 1 and 2 for the moment. Further, $R(.)$ will be positive if and only if $r + b > f'(x)$ so our search will be restricted to intervals where this inequality holds. Assuming only that $f(.)$ and $u(.)$ are twice differentiable, $R(.)$ is differentiable and we have:

$$R'(x) = [r + b - f'(x)]u''(.)[b - f'(x)] - u'(.)f''(x).$$

Proposition 3:
There always exists parameter choices consistent with case 2 such that there are a countable number of candidate steady states in a neighborhood of x_2.

Proof:
In case 2, $r + b - f'(x_2) > 0$ and $bx_2 - f(x_2) > 0$, so $R(x_2) > 0$. Further, since $f''(x_2) < 0$ and $b = f'(x_2)$, we see that $R'(x_2) = -u'(.)f''(x_2) > 0$. The proposition then follows from Lemma 3.   Q.E.D.

Unfortunately, we are unable to get such a clean result for cases 1 and 3. In case 1, the first term in $R'(.)$ is always negative so the multiplicity result will depend on the relative absolute magnitude of u'' and f'. For case 3, x_2 lies in the interval where (8.2) cannot have a solution and again, wherever R is positive and f'' is negative, the first term in R' is negative. Clearly, we could state sufficient conditions on f'' and u'' that would guarantee multiplicities in cases 1 and 3, but we do not do so explicitly here.

To analyze the behavior of dynamics away from steady states, we now employ the standard linearization methodology. It is convenient to revert to the state/costate formulation but defining $Q = -p$ so that all variables take on nonnegative values (recall that since the state variable is a "bad" in this problem, the natural costate is negative). In these units, the equations of motion take the form:

$$dQ/dt = (r + b - f'(x))Q - Bc'(x), \qquad (8.3)$$
$$dx/dt = a(Q) - bx + f(x), \qquad (8.4)$$

where a(Q) is implicitly defined by u'(a) = Q.

The linearized dynamics is defined by the jacobian coefficients:

$$J_{11} = r + b - f'(.); \; J_{12} = -Bc''(.) - u'(.)f''(.); \; J_{21} = 1/u''(.); \; J_{22} = f'(.) - b.$$

Let |J| denote the determinant of J. The eigenvalues are given by solutions to |J−λI| = 0, and solving this quadratic equation, we find:

$$\lambda = [r \pm \sqrt{\{r^2 - 4|J|\}}]/2, \qquad (8.5)$$

where

$$|J| = -r(f' - b) - (b - f') + (1/u'')(Bc'' + u'f'').$$

From this we see that the sign of |J| is critical to local dynamics. If |J| is negative then both eigenvalues are real with one greater than zero and one less than zero. This is the case of saddle value stability where we expect a one dimensional local stable manifold converging to the associated local steady state. On the other hand, if |J| is positive (and r relatively small) then the roots are complex and we expect cyclic dynamics. We will amplify on these possibilities below.

It turns out that for strictly concave u(.), there is a very simple characterization of the sign of |J|. Let the dx/dt = 0 and dQ/dt = 0 curves be represented by $Q_1(x)$ and $Q_2(x)$ respectively. Then we have:

Lemma 4:
Assuming U″ < 0, evaluated at any steady state, Sign {|J|} = Sign {$dQ_1/dx - dQ_2/dx$}.

Proof:
From (8.3) and (8.4) we derive:

$$dQ_1/dx - dQ_2/dx = \qquad (8.6)$$
$$(b - f')u'' - Bc''/(r + b - f') - Bc'f''/[(r + b - f')^2]$$

Consequently, since u″ < 0 and at any steady state, r + b − f' > 0,

$$\text{Sign } \{dQ_1/dx - dQ_2/dx\} = \qquad (8.7)$$
$$\text{Sign } \{-(b - f')(r + b - f') + Bc''/u'' + Bc'f''/(r + b - f')u''\}.$$

Then, substituting for Bc'/(r + b − f') from the steady state condition (8.2) and comparing with the formula for |J|, the lemma follows.   Q.E.D.

From this lemma, we can infer the following:

Proposition 4:
For strictly concave u(.), candidate steady states generically will alternate between two types, type 1 where there will exist a one dimensional local stable manifold converging to the steady state and type 2 which are totally unstable. Further the first and last steady states (assuming there are more than one) are of type 1.

Proof:
Generically (8.6) will be nonzero and the Sign will alternate between positive and negative at successive intersections. Therefore, from Lemma 4, $|J|$ will alternate in sign at successive steady states. When $|J|$ is negative we know we are at a type 1 steady state. When $|J|$ is positive, we can see from (8.5) that either the eigenvalues are real in which case they must both be positive, or they are complex with positive real part – in either case totally unstable. By our assumptions on u(.), f(.) and c(.), $Q_1(0) < \infty$ and $Q_2(0) = \infty$, so the first intersection is of type 1 and since we already established that the number of candidate steady states is generically odd, the last intersection must be of type 1 as well. Q.E.D.

Note that there is some remaining ambiguity concerning the qualitative behavior at type two intersections. If the interest rate is sufficiently small, we can see from (8.5) that eigenvalues will be complex and the dynamics must be an unstable spiral. However, unless r is actually zero, there is the possibility that roots will be real (both positive). We comment on this further when constructing phase diagrams.

The reader might think there could be limit cycles in a neighborhood of the unstable steady states. However, it can be shown using Green's theorem for integration in the plane that such cycles cannot occur in problems of this type as long as r is greater than zero. See for example, Brock and Scheinkman (1977) and Tahvonen and Salo (1996).

We do not attempt a treatment here of the case $r = 0$. For that case, there are open questions about the existence of optima and one must in any case resort to an overtaking criteria since the objective function integral may well be unbounded. There are sufficiency theorems for the overtaking criteria, but they do not apply to nonconvex problems such as ours.

Once there are multiple candidate steady states, there are almost certain to be multiple solutions to the necessary conditions that satisfy transversality, at least for some values of the state variable. Each of these can be thought of as a local maximum (or local minimum) of the value function W(.). The main tool that we have for choosing the global maximum among these is a generalization of that used by Skiba (1978). It is based on use of the Hamilton-Jacobi equation for time autonomous problems:

$$\text{Max}_a\ H(x, Q(x), a) = rW(x), \tag{8.8}$$

where $Q(x)$ represents the optimal stable manifold. Thus, for any candidate locally optimal trajectory $Q_i(x), a_i(x)$, we can associate a candidate value function $V_i(x) = H(x, Q_i(x), A_i(x))/r$ and the global optimum starting at x will be defined by $i(x) = \arg\max V_i(x)$.

Proposition 5: (Candidate Value Function Comparison)[3]
Consider two candidate value functions $V_i(x)$ and $V_j(x)$. Suppose

$$Q_i(x)dx_i(x)/dt \geq Q_j(x)dx_i(x)/dt \qquad (8.9)$$

(which obviously holds if $dx_i/dt = 0$), then $V_j(x) \geq V_i(x)$ with strict inequality if $u(a)$ is strictly concave at a_j or (8.9) holds with strict inequality.

Proof:
Evaluated at any chosen value of the state variable,

$$\begin{aligned} r(V_j - V_i) &= u(a_j) - u(a_i) - Q_j \, dx_j/dt + Q_i \, dx_i dt \\ &\geq u(a_j) - u(a_i) - Q_j \, \{dx_j/dt - dx_i/dt\} \\ &= u(a_j) - u(a_i) - u'(a_j)\{a_j - a_i\} \\ &\geq 0, \end{aligned}$$

where the first inequality follows from assumption (8.9) and the second from concavity of $u(.)$. If (8.9) is strict, the first inequality is strict whereas the second is strict if $u(.)$ is strictly concave at a_j. Q.E.D.

9. Three Candidate Steady States

We know that the simplest generic multiplicity situation will involve three candidate steady states so we take up this situation first. Three candidate steady states can arise in all three of our cases and as in section 5, we label these states α, β and γ. We give a relatively complete discussion of case II first and then indicate modifications that must be made in the other cases. There are two types of ambiguity in the qualitative features of phase diagrams that cannot be resolved without further assumptions on functional forms. The first involves the sign of the slope of the stable eigenspace at the largest steady state and the second involves whether or not the two stable manifolds extend through the entire state space.

The first is illustrated by a comparison of Figures 7a–b which depict local dynamics around the largest equilibrium (we draw these in a, x space for visual convenience). Figure 7a applies when the largest equilibrium occurs in a region where $da/dt = 0$ is decreasing whereas, Figure 7b applies if that equilibrium occurs where it is increasing. The operational distinction between these situations involves whether the optimal loading will be a monotone increasing or decreasing function of the phosphorous stock in this region.

The second ambiguity is illustrated by comparing Figures 8a–c.[4] For Case II we can always associate the lowest equilibrium with an oligotrophic state and the highest with a eutrophic state since the lowest occurs below $\underline{x}$ and the highest above $\bar{x}$. In Figure 8a the oligotrophic stable manifold extends over the entire state space, in Figure 8c, the eutrophic stable manifold extends back to the lowest states and in 8b, both manifolds originate from the unstable equilibrium. Using Proposition 5, we can now state the following:

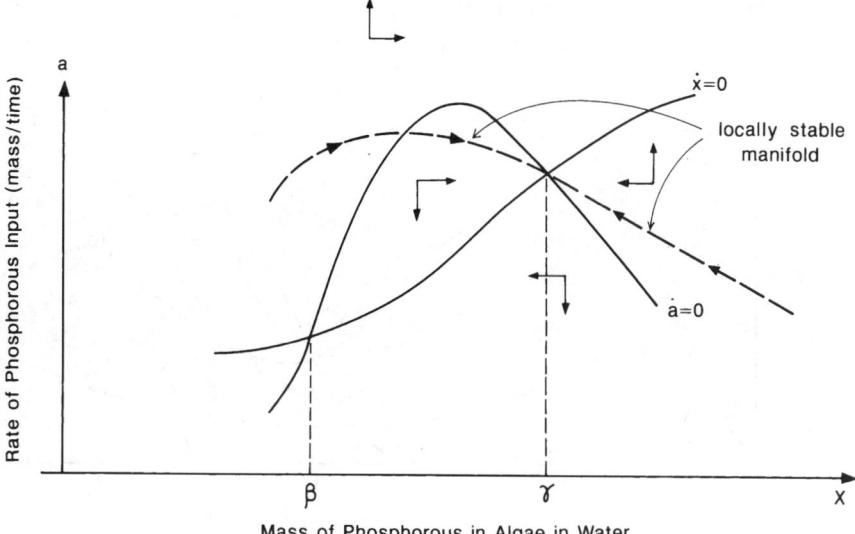

Figure 7a. Phase diagram: à = 0 curve decreasing at steady state.

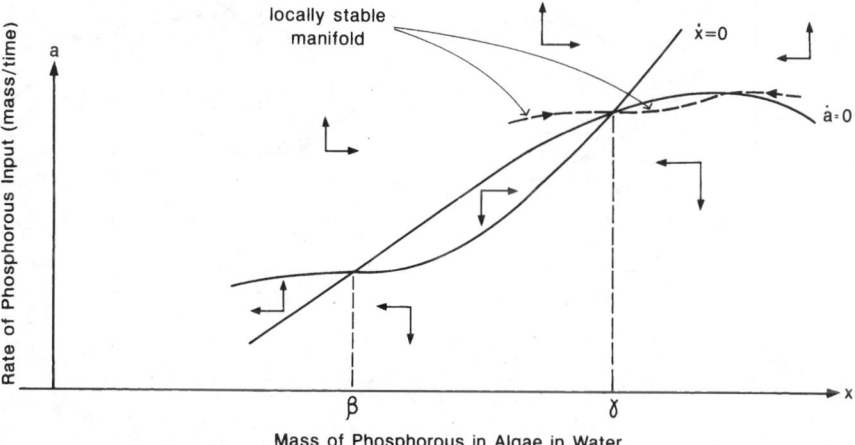

Figure 7b. Phase diagram: à = 0 curve increasing at steady state.

Proposition 6:

(1) If Figure 8a applies, the oligotrophic state is long run optimal from any starting state. (2) If Figure 8c applies, the eutrophic state is long run optimal from any starting state. (3) If Figure 8b applies, there is a Skiba point (S) in the interval [.] such that the oligotrophic state is long run optimal from starting points below S and the eutrophic state is optimal for starting points above.

Proof:

To establish (1) apply proposition 5 at any state greater than or equal to γ with i

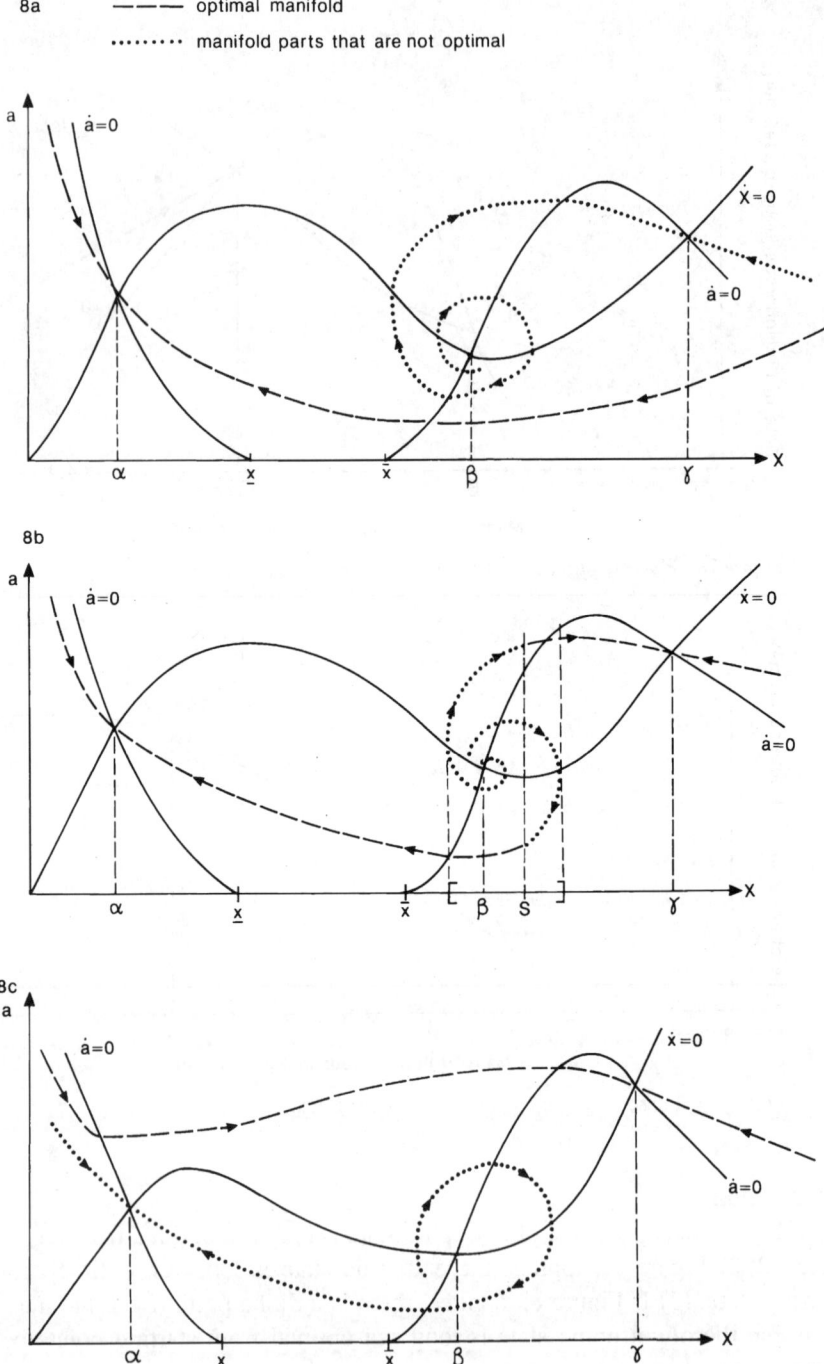

Figure 8. 8a: Phase diagram: oligotrophic steady state α, is optimal. 8b: Phase diagram: oligotrophic steady state α is optimal for initial X_0 below S; γ is optional for initial X_0 greater than S. 8c: Phase diagram: eutrophic steady state γ, is optimal.

representing the eutrophic manifold and j the oligotrophic manifold ($Q_i < Q_j$ and $dx_i/dt \leq 0$) and use connectedness property of optimal paths. To establish (2) apply the proposition to any state less than or equal to α with the reverse orientation. Statement (3) is obvious since the oligotrophic manifold must be optimal to the left of [.], the eutrophic manifold to the right and there can be only one switch point. Q.E.D.

Actually, we can show that for Figure 8b, the Skiba point is strictly inside the interval [.] if u(.) is strictly concave. This is because we can use Proposition 5 in the same way as above to show that the oligotrophic manifold is strictly optimal at the left point of the interval and the eutrophic manifold similarly strictly optimal at the right point.

Obviously, it is quite important for policy to know which of these three situations apply. We can say using a continuity argument that if B is sufficiently close to B* (from below) then situation 8a must apply since at B* the oligotrophic manifold is strictly optimal. Beyond this, we are unable to say anything definitive without knowing more about functional form.

We comment briefly on modifications for cases I and III. The story is pretty much unchanged for case I except that there is no obvious way of locating the positions of α and γ relative to x_m, except for B close to B* where again a continuity argument will guarantee that α is less than x_m. For case III, the $dx/dt = 0$ curve cuts the horizontal axis so situation 8a cannot hold. This makes sense since if the phosphorous level is sufficiently high we know there is no policy that will return us to the oligotrophic state. Generically, there is no unstable steady state since both the $dx/dt = 0$ and the $da/dt = 0$ curves cut the horizontal axis in the region where it would ordinarily be found. This leaves the question of if and where we will find a Skiba point in situations where there are two locally stable equilibria. It turns out that this point (if it exists) is always strictly interior to the region where return to the clean steady state is possible, implying that if we are sufficiently close to the boundary, it is optimal to cross the point of no return and converge ultimately to the dirty steady state. This can be seen with the aid of Figure 9, where the situation with two locally stable steady states is drawn.

The locally stable clean steady state manifold must cut the X axis to the left of breakeven point $\hat{x}$ since it would have to be locally vertical anywhere it crossed the $dx/dt = 0$ line. Then, since an optimal path must exist from any starting state, the other locally stable manifold must extend at least that far into the region to the left of $\hat{x}$. As drawn, there is a Skiba interval and the Skiba point must lie somewhere in the interior of that interval. The reader should also see that it is possible to draw the diagram so that the dirty steady state is the optimal destination starting from any initial stock (so there would not be a Skiba point). Presumably this last situation will apply when B is sufficiently small so that relatively little weight is put on the preferences of enjoyers.

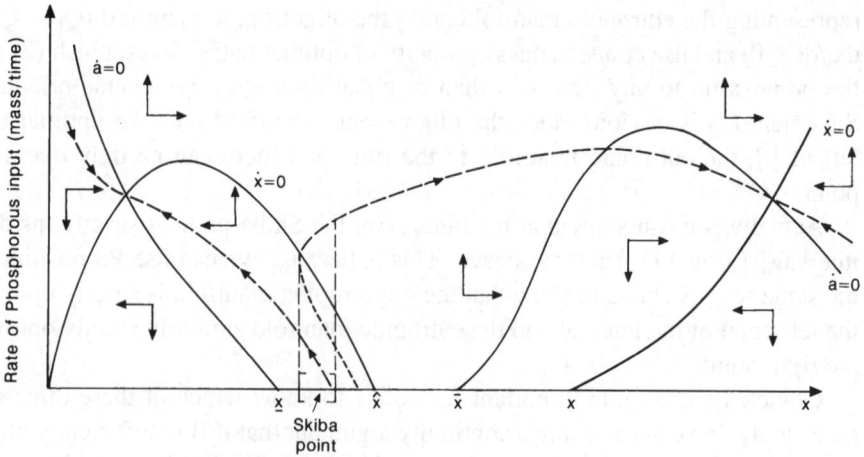

Figure 9. Phase diagram: Case 3 with two steady states, one Skiba point.

10. Five or More Candidate Steady States

The situation with more than three multiplicities and be analyzed in much the same way as when there are three, using proposition 5. In a sense (except in case III), all possibilities can occur. That is (for five candidate steady states) there can be zero, one or two Skiba points. When there are zero, any of the three locally stable manifolds can be the global optimum and when there is one, it can occur between the first and second stable candidates or between the second and third and any pair of manifolds can have basins of attraction. We illustrate in Figures 10a–c by drawing the dynamics (this time for case I) in which the middle locally optimal steady state is always long run optimal (10a), where the first and third have basins of attraction (10b) and where all locally stable steady states have a basin of attraction (10c). We leave it to the reader to fill in other cases.

One could proceed formally by induction to analyze seven or more candidate steady states and show that again, all combinations remain possible but we leave this exercise to the sufficiently interested reader.

11. Conclusions

In this paper, we have characterized the types of behavior that can be optimal in the lake model of Carpenter et al. (1999) and shown how these behaviors change as functions of key parameters such as the relative weight put on affected groups. However, particularly in cases where there are multiple candidate steady states we are unable to resolve some key ambiguities such as the existence and location of Skiba points and the resulting sizes of basins of attraction. Unfortunately, these ambiguities seem to be characteristic of the class of nonconvex problems of which ours is an example. When there are multiple local optima, it is rarely possible to choose the best among them except by direct numerical comparison.

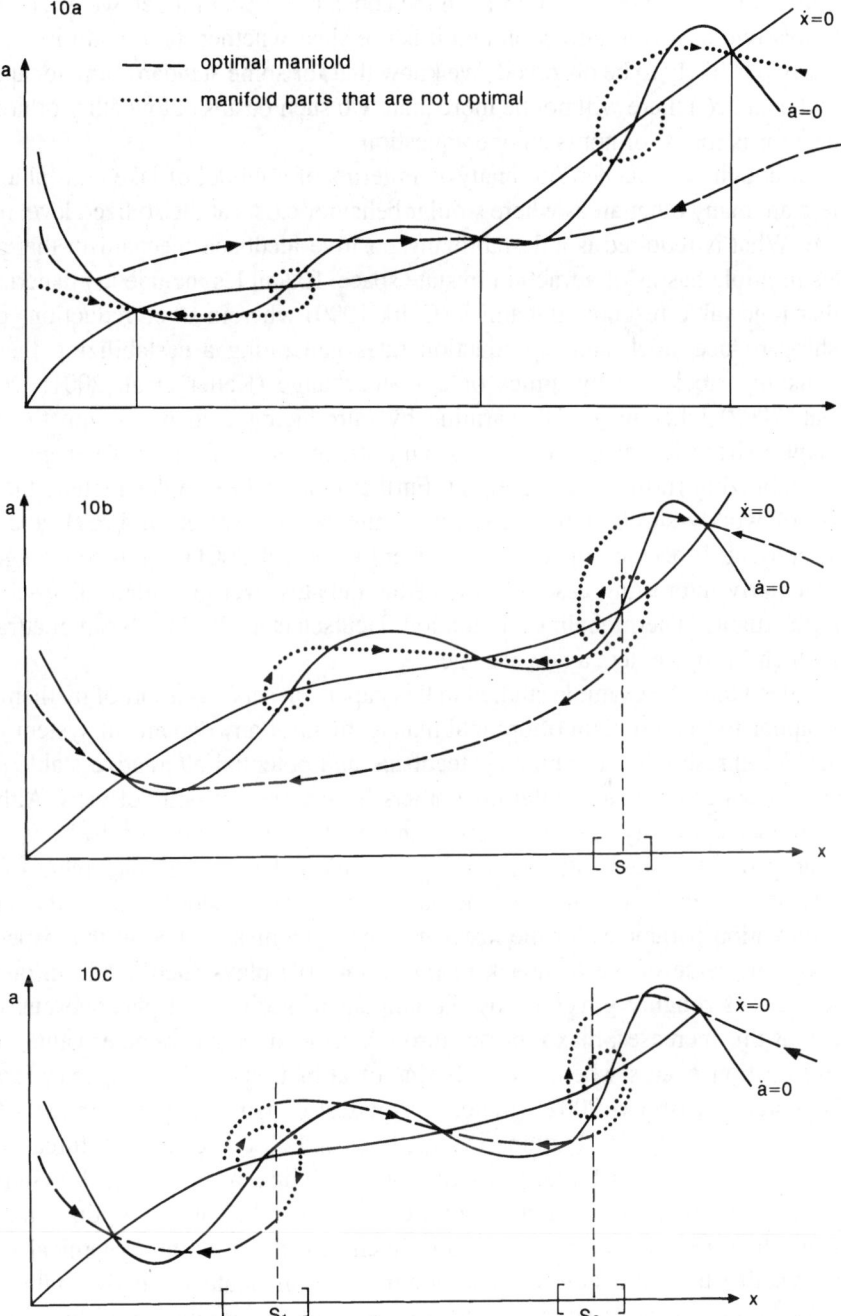

Figure 10. 10a: Phase diagram: five steady states, middle one is optimal. 10b: Phase diagram: five steady states, one Skiba point. 10c: Phase diagram: five steady states, two Skiba points.

A second ambiguity that we have uncovered involves the number of candidate steady states. Drawing intuition from the constant loading model, we might expect at most two basins of attraction and it is not clear whether the possibility of more than two is likely to be observed. We know that for some standard functional forms for f(.) and c(.) there will not be more than two such basins but whether or not such behavior is robust or not is an open question.

Although we couched our analysis in terms of a model of lake eutrophication, there are many other areas where similar behavior exists at the stylized level treated here. What is required is a destabilizing positive feedback mechanism that generates multiple basins of attraction in state space. Examples include (1) fisheries and other renewable resource dynamics (Clark 1990) wherein stock reductions due to fishing reduce stock and reproduction rates generating a destabilizing feedback on fishing stock, (2) dynamics of climate change (Keller et al. 2001; Deutsch et al. 2002) wherein global warming by introducing cold water into the North Atlantic (from ice melt) inhibits the gulf stream flow of warm water generating a destabilizing further cooling effect. Further potential examples include the invasion of woody shrubs into grassland savannahs (Walker et al. 1969), and other eutrophying processes in fjords and rivers (Naevdal 2001). Climate change is a particularly interesting case in view of the debates over potential collapse of the North Atlantic Thermohaline Circulation (Deutsch et al. 2001), a basin of attraction in which Europe enters a new ice age!

Indeed the lake example studied in this paper can serve as a sort of mathematical metaphor for the problem of optimal management of any dynamical system where there are thresholds, destabilizing feedback and potential alternative stable states, some of which are much better than others from a welfare point of view. Although our model is too stylized to describe any of these situations precisely, we think it can provide insight into expected qualitative behaviors. Management of such systems becomes especially treacherous when there is a slow variable that acts as a bifurcation parameter for the faster moving dynamics studied in this paper. The maximum value of the feedback response rate (f') plays such a role in our lake model. It is roughly governed by the amount of sedimented phosphorous in the lake; as this increases, max f' passes through the cleansing parameter value (b) and a bifurcation occurs wherein new basins of attraction suddenly appear. Ludwig, Carpenter and Brock (2002) studied a slow/fast dynamical system model of lake eutrophification by mostly numerical methods and observed such a bifurcation. An important problem for future research to get analytical results for such systems.

Another important problem for future research is the study of optimal management when there is uncertainty about which model drives the dynamical system and whether there is a slow variable present that can create alternative stable states in the faster dynamics acting at the frequency typically considered by managers driven by economic incentives.

12. Acknowledgments

Brock thanks Dee Dechert for many helpful conversations on this paper. Brock would like to thank the NSF under Grant # SES-9122344 and the Vilas Trust for financial support. Starrett thanks Karl-Göran Mäler for helpful discussions. None of the above are responsible for errors in this paper.

Notes

1. It is interesting to note that a concave version of this problem for which a sufficiency theorem holds requires that f be convex rather than concave. The reason is that the state variable here is a "bad" which reverses sign conventions. Of course, when this theorem holds there can be only one local optimum satisfying transversality and indeed with f(.) convex, there is only one solution to (5.10). However, the nonconvex element is fundamental to our problem and a concave segment of f(.) is well established empirically.
2. This must be possible since we assumed that $f'(x)$ goes to zero in x.
3. This proposition can be generalized; it holds for any time autonomous control problem for which the accumulation equation is linear and separable in the control variable, and the felicity function is concave in the control for each fixed value of the state variable.
4. We have drawn these consistent with Figure 7a but the same analysis applies to Figure 7b.

References

Becker, R. and J. Boyd (1997), *Capital Theory, Equilibrium Analysis, and Recursive Utility*. Oxford: Basil Blackwell.

Brock, W. (1970), 'On Existence of Weakly Maximal Programmes in a Multi-Sector Economy', *Review of Economic Studies* **37**, 275–280.

Brock, W. (1977), 'A Polluted Golden Age', in V. Smith, ed., *Economics of Natural and Environmental Resources*. New York: Gordon and Breach.

Brock, W. and W. Dechert (1983), *The Generalized Maximum Principle*. SSRI Working Paper, # 8316. Madison: Department of Economics, University of Wisconsin.

Brock, W. and W. Dechert (1985), 'Dynamic Ramsey Pricing', *International Economic Review* **26**, 569–591.

Brock, W. and A. Haurie (1976), 'On Existence of Overtaking Optimal Trajectories Over an Infinite Time Horizon', *Mathematics of Operations Research* **1**, 337–346.

Brock, W. and A. Malliaris (1989), *Differential Equations, Stability and Chaos in Dynamic Economics*. Amsterdam: North Holland.

Brock, W. and J. Scheinkman (1977), 'Global Asymptotic Stability of Optimal Control', in J. Pitchford and S. Turnovsky, eds., *Applications of Control Theory to Economic Analysis* (pp. 173–205). Amsterdam: North-Holland.

Carlson, D., A. Haurie and A. Leizarowitz (1991), *Infinite Horizon Optimal Control*. Berlin: Springer-Verlag.

Carpenter, S., D. Ludwig and W. Brock (1999), 'Management of Eutrophication for Lakes Subject to Potentially Irreversible Change', *Ecological Applications* **9**(3), 751–771.

Clark, C. (1990), *Mathematical Bioeconomics: The Optimal Management of Renewable Resources*, Second Edition. New York: John Wiley and Sons.

Deutsch, C., M. Hall, D. Bradford and K. Keller (2002), *Detecting a Potential Collapse of the North Atlantic Thermohaline Circulation: Implications for the Design of an Ocean Observation*

System. Program in Atmospheric and Oceanic Sciences and Department of Economics, Princeton University; Department of Geosciences, Penn State University.

Ekelund, I. and J. Scheinkman (1986), 'Transversality Conditions for Some Infinite Horizon Discrete Time Optimization Problems', *Mathematics of operations research* **13**, 216–229.

Hartman, P. (1973), *Ordinary Differntial Equations*. New York: John Wiley and Sons.

Hochman, O. and E. Hochman (1980), 'Regeneration, Public Goods, and Economic Growth', *Econometrica* (July) **48**(5), 1233–1250.

Keller, K., B. Bolker and D. Bradford (2001), *Uncertain Climate Thresholds and Optimal Economic Growth*. Department of Geosciences, Penn State University; Department of Zoology, University of Florida; Department of Economics, Princeton University.

Ludwig, D., S. Carpenter and W. Brock (2002), 'Optimal Phosphorus Loading for a Potentially Eutrophic Lake', *Ecological Applications* (forthcoming).

Naevdal, E. (2001), 'Optimal Regulation of Eutrophying Lakes, Fjords, and Rivers in the Presence of Threshold Effects', *American Journal of Agricultural Economics* (November) **83**(4), 972–984.

Scheffer, M. (1997), *The Ecology of Shallow Lakes*. New York: Chapman and Hall.

Scheffer, M., W. Brock and F. Westley (2000), 'Socioeconomic Mechanisms Preventing Optimum Use of Ecosystem Services: An Interdisciplinary Theoretical Analysis', *Ecosystems* **3**, 451–471.

Skiba, A. (1978), 'Optimal Growth with a Convex-Concave Production Function', *Econometrica* (May) **46**(3), 527–539.

Spence, M. and D. Starrett (1975), 'Most Rapid Approach Paths in Accumulation Problems', *International Economic Review* (June) **16**, 388–403.

Tahvonen, O. and S. Salo (1996), 'Nonconvexities in Optimal Pollution Accumulation', *Journal of Environmental Economics and Management* **31**, 160–177.

Walker, B. H., D. Ludwig, C. Holling and R. Peterman (1969), 'Stability of Semi-arid Savanna Grazing Systems', *Ecology* **69**, 473–498.

The Economics of Shallow Lakes *

KARL-GÖRAN MÄLER[1], ANASTASIOS XEPAPADEAS[2] and
AART DE ZEEUW[3],*
[1]*The Beijer International Institute of Ecological Economics;* [2]*Department of Economics, University of Crete;* [3]*Department of Economics and CentER, Tilburg University, The Netherlands, E-mail: a.j.dezeeuw@uvt.nl (*Author for correspondence)*

Abstract. Ecological systems such as shallow lakes are usually non-linear and display discontinuities and hysteresis in their behaviour. These systems often also provide conflicting services as a resource and a waste sink. This implies that the economic analysis of these systems requires to solve a non-standard optimal control problem or, in case of a common property resource, a non-standard differential game. This paper provides the optimal management solution and the open-loop Nash equilibrium for a dynamic economic analysis of the model for a shallow lake. It also investigates whether it is possible to induce optimal management in case of common use of the lake, by means of a tax. Finally, some remarks are made on the feedback Nash equilibrium.

Key words: ecological systems, non-linear differential games

1. Introduction

The purpose of this paper is to develop an cconomic analysis of the shallow lake. Lakes have been studied intensively and the shallow lake model is well tested and documented, so that the analysis has a direct meaning. However, the lake model can also be viewed as a metaphor for many of the ecological problems facing mankind today, so that the analysis developed in this paper will have a wider applicability. The economic analysis is especially challenging because of the non-linear dynamics of the lake (which yields non-convex decision problems) and the gaming aspects related to the common property character of the lake.

It has been observed that shallow lakes, due to a heavy use of fertilizers on surrounding land and an increased inflow of waste water from human settlements and industries, at some point tend to flip from a clear state to a turbid state with a greenish look caused by a dominance of phytoplankton (Carpenter and Cottingham 1997; Scheffer 1997). The release of nutrients, especially phosphorus, into the lake stimulates the growth of phytoplankton and in addition to that, the resulting turbidity prevents light to reach the bottom of the lake so that submerged vegetation

---

* This research was initiated at a meeting of the Resilience Network which had financial support from the MacArthur Foundation. We are very grateful for the advice and comments of (in alphabetical order) William Brock, Steve Carpenter, Davis Dechert, Marten Scheffer, Perry Shapiro, Sjak Smulders, Robert Solow, David Starrett, Florian Wagener and the referees.

disappears. With the vegetation many species disappear such as waterfleas which graze on phytoplankton. It has also been observed that shallow lakes are hard to restore in the sense that the nutrient loads have to be reduced below the level where the flip occurred before the lake flips back to a clear state. The positive feedback through the effect on the submerged vegetation is one explanation for this so-called hysteresis effect.

Ecological systems often display discontinuities in the equilibrium states of the system over time. A seminal paper in this area models the sudden outbreak of an insect, called the spruce budworm, and the long time it takes before the budworm density jumps back to a low number again (Ludwig, Jones and Holling 1978). Technically, this hysteresis effect can be modelled by a non-linear differential equation which has multiple steady-states with separated domains of attraction in a certain range of the exogenous variable. Other examples of ecological systems with hysteresis, among which the lake model, are described in Ludwig, Walker and Holling (1997).

In the ecological literature, management of shallow lakes is mostly interpreted as preventing the lake to flip or, if it flips, as restoring the lake in its original state. However, this approach denies the economics of the problem in the sense of the trade-offs between the utility of the agricultural activities, which are responsible for the release of phosphorus, and the utility of a clear lake. When the lake flips to a green turbid state, the value of the ecological services of the lake (e.g., the intake of water and recreation) decreases but this situation corresponds to a high level of agricultural activities. It depends, of course, on the relative weight attached to these welfare components whether it is better to keep the lake clear or not. Note that if it is better to keep the lake clear, it is very costly to let the lake flip first because of the hysteresis. A second economic issue is that lakes are usually common property resources and therefore suffer from sub-optimal use, in the absence of coordination.

The literature on the lake model is rapidly increasing. Carpenter, Ludwig and Brock (1999) focus on hysteresis and irreversibility issues. The paper that comes closest to this one is by Brock and Starrett (1999). They consider the dynamics and the optimal management of the lake and point out the occurrence of saddle-point stable steady-states and Skiba points. This paper extends their analysis to Nash equilibria, for the game of common property, and to tax policies with the aim to internalize the externalities (see below). Brock and de Zeeuw (2002) consider a repeated game version of the lake model. They show that the occurrence of "bad" Nash equilibria can in fact be beneficial because with these points as threats in trigger strategies, cooperation can be sustained for lower values of the discount factor.

In the first part of the paper, very simple welfare analysis is done on the possible steady-states of the lake model. Relative weights are chosen such that it is optimal to manage the lake in one of its clear states, called oligotrophic states. It is shown, however, that when the lake is shared by more than one community, two Nash equilibria occur: one in an oligotrophic state and one in a dirty state, called a eutrophic state. In the second part of the paper, intertemporal welfare is maximized subject to the dynamics of the lake. It is shown that in case the discount rate is low enough, an optimal path for phosphorus loadings exists, from each initial condition of the lake, which moves the lake towards its optimal steady-state. When the lake is shared by more than one community, a non-linear differential game has to be solved. The phase-diagram for the open-loop Nash equilibrium has three steady-states, two of which are saddle-point stable and correspond to the Nash equilibria found in the first part of the paper. The third point is unstable and displays complex dynamics. However, it is shown that a so-called Skiba point exists which splits the possible initial conditions of the lake in an area from where the Nash equilibrium loading trajectory will approach the oligotrophic saddle-point, and an area from where the eutrophic saddle-point results.

The question arises whether it is possible, by levying a tax on the loading of phosphorus, to induce the communities to follow an optimal management path. Note that if the communities are locked in the eutrophic Nash equilibrium, a straight path to the optimal steady-state is not feasible due to the hysteresis. Assuming that it is not possible to implement a time-varying tax, the answer depends on the number of communities. It is shown that if the number is low enough, a constant tax yields a Nash equilibrium path that moves towards the optimal steady-state (although this path will not be the same as the optimal management path, of course). If the number is high, however, more saddle-points arise again and the dynamics becomes very complex, so that there is no guarantee that a constant tax can induce optimal management of the lake in the long run.

A final issue regards the type of Nash equilibrium employed in the analysis. It is well-known that the open-loop Nash equilibrium is not strongly time-consistent and therefore a feedback Nash equilibrium is preferred. However, due to the non-linear dynamics of the lake, it is very difficult to find a feedback Nash equilibrium. In the last section of the paper, some preliminary remarks are made on this issue. The problem is a one-dimensional infinite horizon differential game, so that the techniques developed by Tsutsui and Mino (1990) for dynamic duopolies with sticky prices, may apply. This would imply the occurrence of multiple equilibria, possibly with welfare levels close to optimal management. The complete analysis is left for further research.

The paper is organized as follows. Section 2 describes the shallow lake model. Section 3 is concerned with the economics of the lake steady-states and section 4 with the dynamic welfare analysis of the lake. Section 4 contains the case of optimal management, the open-loop Nash equilibrium, the effect of taxes and the feedback Nash equilibrium. Section 5 concludes the paper.

2. The Lake Model

Shallow lakes have been studied intensively over the last two decades and it has been shown that the essential dynamics of the eutrophication process can be modelled by the differential equation

$$\dot{P}(t) = L(t) - sP(t) + r\frac{P^2(t)}{P^2(t) + m^2}, \quad P(0) = P_0, \tag{1}$$

where P is the amount of phosphorus in algae, L is the input of phosphorus (the "loading"), s is the rate of loss consisting of sedimentation, outflow and sequestration in other biomass, r is the maximum rate of internal loading and m is the anoxic level (see for an extensive treatment of the lake model Carpenter and Cottingham 1997 or Scheffer 1997). Less is known about deep lakes but from what is known now, it can be expected that the same type of model will be adequate. However, estimates of the parameters of this differential equation for different lakes vary considerably, so that a wide range of possible values has to be considered.

By substituting $x = P/m, a = L/r, b = sm/r$ and by changing the time scale to rt/m, equation (1) can be rewritten as

$$\dot{x}(t) = a(t) - bx(t) + \frac{x^2(t)}{x^2(t) + 1}, \quad x(0) = x_0. \tag{2}$$

In order to understand some of the important features of this model, suppose that the loading a is constant. What happens depends on the value of the parameter b. If $b \geq 3\sqrt{3}/8$, all values of a lead to one stable steady-state (see Figure 1). If $b \leq 1/2$, values of a above the local maximum of the curve of steady-states in Figure 2 lead to one stable steady-state again. However, values of a below this local maximum yield two stable steady-states for the differential equation (2). The domains of attraction are determined by the unstable steady-state in between: to the right of this point the high stable steady-state results and to the left the low one. If $1/2 < b < 3\sqrt{3}/8$, values of a below the local minimum and above the local maximum of the curve of steady-states in Figure 3 lead to one stable steady-state. For values of a in between, two stable steady-states occur again for the differential equation (2), with domains of attraction divided by the unstable steady-state.

It is easy to see a hysteresis effect now for $b < 3\sqrt{3}/8$. If the loading a is gradually increased, at first the steady-state level of phosphorus remains low: the lake remains in an oligotrophic state with a high level of ecological services. At a certain point, however, the lake *flips* to a high steady-state level of phosphorus. To put it differently, the lake *flips* to a eutrophic state with a low level of ecological services. If it is then decided to lower the loading a in order to try to bring the lake back to an oligotrophic state, it is not enough to reduce a just below that flip-point. If b is high enough ($1/2 < b < 3\sqrt{3}/8$, Figure 3), it can still be done, but a has to be reduced further until the lake flips back to an oligotrophic state. If $b \leq 1/2$ (Figure 2), however, then the lake is trapped in high steady-state levels

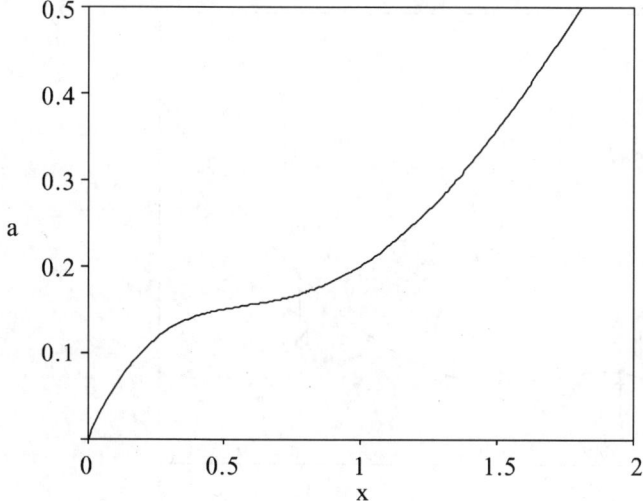

Figure 1. Lake equilibria (b = 0.7).

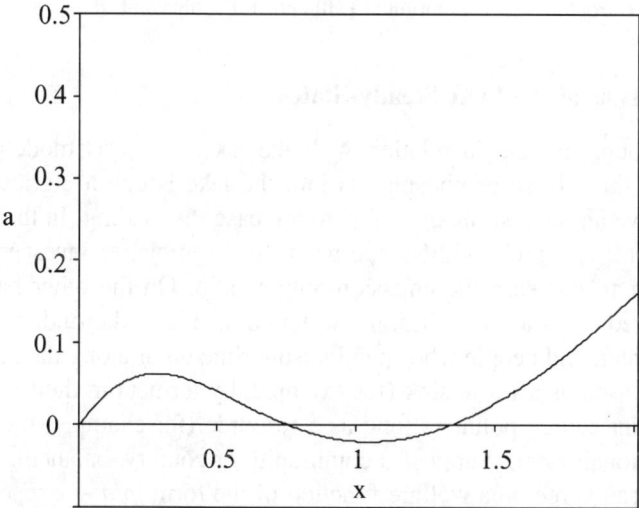

Figure 2. Hysteresis, irreversible (b = 0.48).

of phosphorus which means that the first flip is irreversible. In that case, only a change in the parameter b (e.g., by releasing a certain type of fish and thus changing the fauna) can restore the lake. In the sequel of the paper, it is assumed that the parameter $b = 0.6$ so that the lake displays hysteresis but a flip to a eutrophic state is reversible. Furthermore, the loading a will not be exogenous anymore but subject to control. In section 3, a is still constant and the trade-off is considered between the benefits of being able to release that constant amount of phosphorus, on the one hand, and the resulting damage to the lake, on the other hand. Section 4 provides a full dynamic analysis where a can change over time.

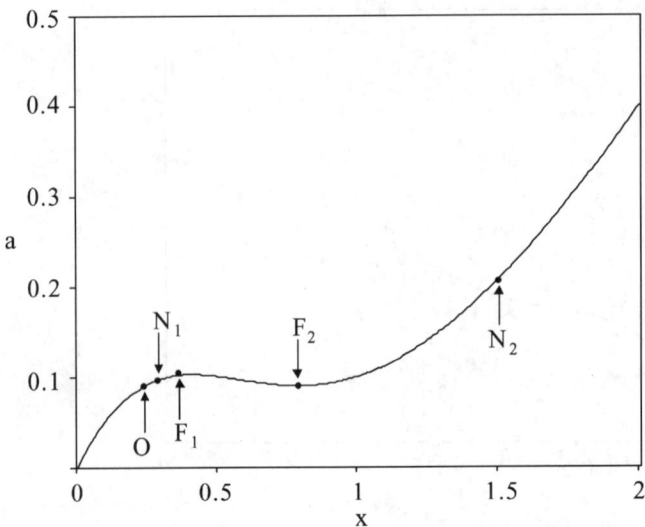

Figure 3. Hysteresis, reversible (b = 0.6). O: optimal management; N_1: oligotrophic Nash equilibrium; N_2: eutrophic Nash equilibrium; F_1: flip point; F_2: flip point.

3. Economic Analysis of the Lake Steady-States

Several interest groups operate in relation with the lake, that was modelled in section 2. Because the release of phosphorus into the lake is due to agricultural activity, farmers have an interest in being able to increase the loading. In that way, the agricultural sector can grow without the need, for example, to invest in new technology in order to decrease the emission-output ratio. On the other hand, a clean lake is preferred by fishermen, drinking water companies, other industry that makes use of the water, and people who spend leisure time on or along the lake. In general, the lake is used as a waste sink (for example, by farmers in their activity as multiple non-point source polluters) and as a resource (for example, by water utilities and recreational users). Suppose a community or country, balancing these different interests, can agree on a welfare function of the form $\ln a - cx^2, c > 0$. The lake has value as a waste sink for agriculture ($\ln a$), for example, and it provides ecological services that decrease with the amount of phosphorus ($-cx^2$). The parameter c reflects the relative weight of these welfare components. Suppose, furthermore, that the lake is shared by n communities or countries with the same welfare function. In this section it is assumed that the communities choose constant loading levels $a_i, i = 1, \ldots, n$, and that the amount of phosphorus adjusts instantaneously to its steady-state level. A logarithmic functional form for the welfare function is chosen, because it is convenient for the technicalities of the analysis and because the optimal management outcome in terms of total loading will be independent of the number of communities. This is helpful because the number of communities can be varied while the optimal management outcome as a benchmark remains the same. It is assumed that the area around the lake is large enough so that adding new communities does not lead to crowding out: the objectives are assumed to be additive in the number n.

Optimal management of the lake requires to solve

$$\text{maximize} \sum_{i=1}^{n} \ln a_i - ncx^2 \quad \text{s.t.} \quad a - bx + \frac{x^2}{x^2+1} = 0, a = \sum_{i=1}^{n} a_i. \quad (3)$$

Simple calculus shows that the optimal amount of phosphorus is determined by

$$b - \frac{2x}{(x^2+1)^2} - 2cx(bx - \frac{x^2}{x^2+1}) = 0. \quad (4)$$

Optimal management, of course, does not necessarily yield an oligotrophic state for the lake. If the communities attach a relatively low weight c to ecological services, it can be optimal to choose a eutrophic state with a high level of agricultural activities. It is easy to show that for large values of c, the optimal management problem has one maximum for an x below the flip-point. As the value of c is decreased, first a local maximum appears for a high x whereas the global maximum is still achieved for a low x, but for c low enough (i.e., $c \leq 0.36$) the global maximum occurs for a high x beyond the flip-point. In the sequel of the paper it is assumed that enough weight (i.e., $c = 1$) is attached to the services of the lake to make it optimal to aim for an oligotrophic state.

If $c = 1$, equation (4) yields $x^* = 0.33$ with total loading $a^* = 0.1$. Note that the same level of total loading can also lead to the eutrophic state $x = 1$, if the initial amount of phosphorus is in the upper domain of attraction (see Figure 3). A flip occurs when total loading is increased to $a = 0.1021$, so that the lake is managed not far from what is called the "edge of hysteresis" (Brock, Carpenter and Ludwig 1997). A small mistake may cause a flip with high costs, not only directly because of a jump to a high x but also indirectly because of the long return path.

If the communities do not cooperate in managing the lake, it is assumed a Nash equilibrium results which requires to solve

$$\text{maximize} \ln a_i - cx^2, i = 1, \ldots, n, \quad \text{s.t.} \quad a - bx + \frac{x^2}{x^2+1} = 0, \quad (5)$$

$$a = \sum_{i=1}^{n} a_i.$$

Simple calculus shows that the Nash equilibrium level of phosphorus is determined by

$$b - \frac{2x}{(x^2+1)^2} - 2\frac{c}{n}x(bx - \frac{x^2}{x^2+1}) = 0. \quad (6)$$

If $c = 1$ again and if the number of communities $n = 2$, equation (6) has three solutions, two of which correspond to a Nash equilibrium. The first Nash equilibrium yields $x^N{}_1 = 0.36$ with total loading $a^N{}_1 = 0.1012$. The lake is still in an oligotrophic state but closer to the edge of hysteresis. However, the second Nash equilibrium yields an eutrophic state $x^N{}_2 = 1.51$ with total loading $a^N{}_2 = 0.2108$. Welfare under optimal management and in the oligotrophic Nash equilibrium are comparable, but welfare in the eutrophic Nash equilibrium is much lower. Moreover, when the communities are locked into the second Nash equilibrium and decide to coordinate, it is much more difficult to reach the optimal management outcome, due to the hysteresis. It is not enough to reduce total loading to 0.1. It has to be reduced to 0.0898 first, in order to flip back to an oligotrophic state, and can then be increased to 0.1 again.

If $n > 2$, these numbers change of course, but it is easy to see that for all b in the range with hysteresis and reversibility ($1/2 < b < 3\sqrt{3}/8$), on which this paper focuses, always two Nash equilibria occur. In fact, equation (6) intersects the curve for the lake steady-states with the curve described by $(n/2cx)(b - 2x/(x^2 + 1)^2)$. For b in the range given above, this curve has a negative part for x in a positive range. Furthermore, it approaches infinity for $x \downarrow 0$ and it approaches zero from above for $x \to \infty$. Increasing the number of communities n implies that the curve is stretched out but the three intersection points remain, two of which are Nash equilibria.

In the next section the loading a can change over time and the amount of phosphorus does not adjust instantaneously to its steady-state level but gradually according to equation (2), which turns the optimal management problem into an optimal control problem and the static game into a differential game. The Nash solutions found in this section return (approximately) as saddle-point stable steady-states with solution trajectories that may have to bend around the flip-point.

4. Dynamic Economic Analysis of the Lake Model

Suppose that the problem has an infinite horizon, so that the objectives become

$$W_i = \int_0^\infty e^{-\rho t}[\ln a_i(t) - cx^2(t)]dt, \quad i = 1, \ldots, n, \tag{7}$$

where $\rho > 0$ is the discount rate.

4.1. OPTIMAL MANAGEMENT

Optimal management requires to maximize the sum of the objectives W_i, subject to equation (2) with $a = \sum a_i$. This is an optimal control problem and the maximum principle yields the necessary conditions

$$\frac{1}{a_i(t)} + \lambda(t) = 0, \quad i = 1, \ldots, n, \tag{8}$$

$$\dot{\lambda}(t) = [(b + \rho) - \frac{2x(t)}{(x^2(t) + 1)^2}]\lambda(t) + 2ncx(t), \tag{9}$$

with a transversality condition on the co-state λ, and equation (2). Using (8), equation (9) can be rewritten as a set of identical differential equations in a_i, $i = 1, \ldots, n$. The sum of these equations (or multiplication of one of them by n) yields a differential equation in total loading a:

$$\dot{a}(t) = -[(b+\rho) - \frac{2x(t)}{(x^2(t)+1)^2}]a(t) + 2cx(t)a^2(t). \tag{10}$$

The solution is given by the set of differential equations (2) and (10), and the transversality condition. Note that $b = 0.6$ (see section 2) and $c = 1$ (see section 3). The phase-diagram in the (x, a)-plane is drawn in Figure 4a. One curve represents the steady-states for x and can be recognized as the lake steady-states, which were discussed in sections 2 and 3. The other curve represents the steady-states for a. Its position depends on the discount rate ρ. If the discount rate is low enough ($\rho < 0.1$), this curve intersects the first curve only once in a point that is saddle-point stable. If the discount rate is higher, the second curve moves up, it intersects the first curve three times, and the analysis becomes similar to the analysis of the open-loop Nash equilibrium below. It is assumed here that the discount rate $\rho = 0.03$, which yields the graph in Figure 4a. The steady-state is close to the static optimal-management solution in section 3, and converges to that point when the discount rate goes to 0. The optimal solution prescribes to jump, at any initial state of the lake, to the stable manifold and to move towards the steady-state. Given the non-linearity of the problem, it is not easy to obtain an analytical expression for the stable manifold but a numerical approximation is not difficult to develop. Starting at the steady-state point, the characteristic vector corresponding to the negative eigenvalue of the Jacobian matrix determines the direction of the stable manifold. Working backwards from the steady-state in small steps, a piecewise linear approximation of the stable manifold is then found and the approximation gets better the smaller the steps. With Mathematica (Wolfram 1999), the stable and unstable manifolds for the set of differential equations (2) and (10) can be drawn (see Figure 4b). Note that the stable manifold can be reached from all initial states x_0 and bends around the lower flip-point (see also section 3).

4.2. OPEN-LOOP NASH EQUILIBRIUM

The open-loop Nash equilibrium (Başar and Olsder 1982) is found by applying the maximum principle to each objective W_i, $i = 1, \ldots, n$, separately (fixing a_j for $j \neq i$) subject to equation (2) with $a = \sum a_i$. The set of necessary conditions becomes

$$\frac{1}{a_i(t)} + \lambda_i(t) = 0, i = 1, \ldots, n, \tag{11}$$

$$\dot{\lambda}_i(t) = [(b+\rho) - \frac{2x(t)}{(x^2(t)+1)^2}]\lambda_i(t) + 2cx(t), i = 1, \ldots, n, \tag{12}$$

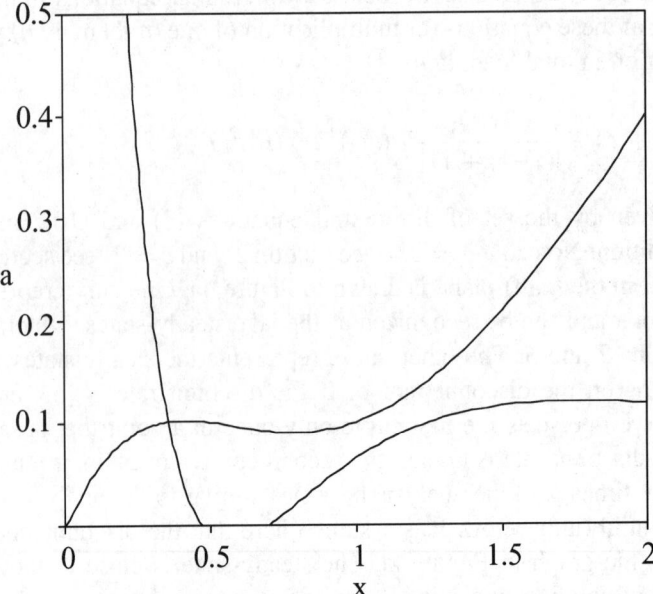

Figure 4a. Phase diagram, optimal management.

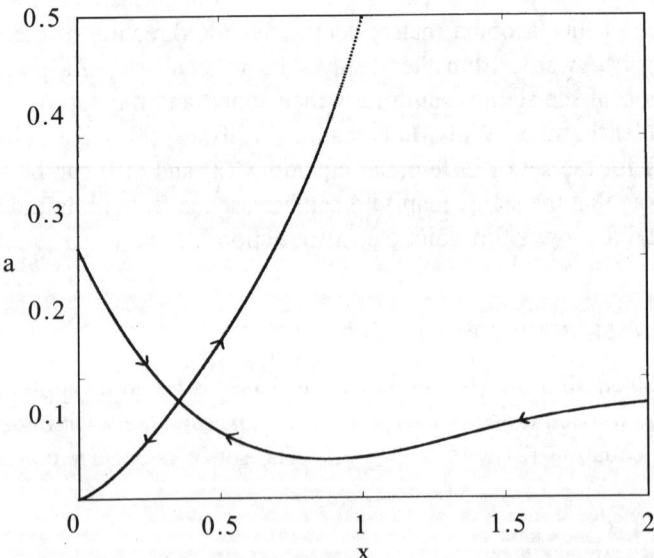

Figure 4b. Stable and unstable manifolds, optimal management.

with transversality conditions on the co-states λ_i, and equation (2). Using (11), equations (12) can be rewritten as differential equations in a_i, $i = 1, \ldots, n$. These equations are identical and the sum (or multiplication of one of them by n) yields a differential equation in total loading a:

$$\dot{a}(t) = -[(b+\rho) - \frac{2x(t)}{(x^2(t)+1)^2}]a(t) + 2\frac{c}{n}x(t)a^2(t). \tag{13}$$

The open-loop Nash equilibrium is given by the set of differential equations (2) and (13), and the transversality conditions. The phase-diagram for two communities $n = 2$ (and $b = 0.6, c = 1, \rho = 0.03$) in the (x, a)-plane is drawn in Figure 5a. The steady-state curves for x and a now have three intersection points. The intersection points on the left and on the right are saddle-point stable and yield possible steady-states for the Nash equilibrium in an oligotrophic and in a eutrophic area, respectively. The intersection point in the middle is unstable with complex eigenvalues. Again with Mathematica (Wolfram 1999), the stable and unstable manifolds for the set of differential equations (2) and (13) can be drawn (see Figure 5b). The trajectories of the stable manifold curl a while from the intersection point in the middle and then go either to the steady-state on the left or to the steady-state on the right. It is clear that when the initial state x_0 lies to the right of the set of curls, the open-loop Nash equilibrium follows the upper trajectory to the steady-state on the right, and when the initial state lies to the left of that area, it follows the lower trajectory to the steady-state on the left. However, it is more difficult to see what happens in the range in between. It can be shown (Appendix A) that a state x_S exists such that for $x_0 < x_S$, the open-loop Nash equilibrium jumps to the lower trajectory and moves towards the oligotrophic steady-state whereas for $x_0 > x_S$, it jumps to the upper trajectory and moves towards the eutrophic steady-state. The point x_S is called a Skiba point because it was introduced by Skiba in an optimal growth model with a convex-concave production function (Skiba 1978; Brock and Malliaris 1989).

If $n > 2$, the same arguments as in section 3 can be used to show that always two open-loop Nash equilibria occur. Note, however, by inspection of equation (13), that the arguments do not hold for all b in the range $(1/2, 3\sqrt{3}/8)$ anymore, because of the positive discount rate ρ, but only for $b + \rho < 3\sqrt{3}/8$, which holds for the specific values chosen for b and ρ.

Before turning to the question whether a tax can induce the communities to choose loadings according to the optimal management trajectory, it is useful to make a few general remarks on the analysis above. First, when comparing equations (10) and (13), it is immediately clear that the open-loop Nash equilibrium also results under optimal management with parameter c/n instead of parameter c. It is an example of a potential game where Nash equilibria can be found by maximizing some adapted objective (Monderer and Shapley 1996; Dechert and Brock 1999). Second, it also means that all outcomes considered here (optimal management with varying relative weight c, and symmetric open-loop Nash equilibria with fixed c

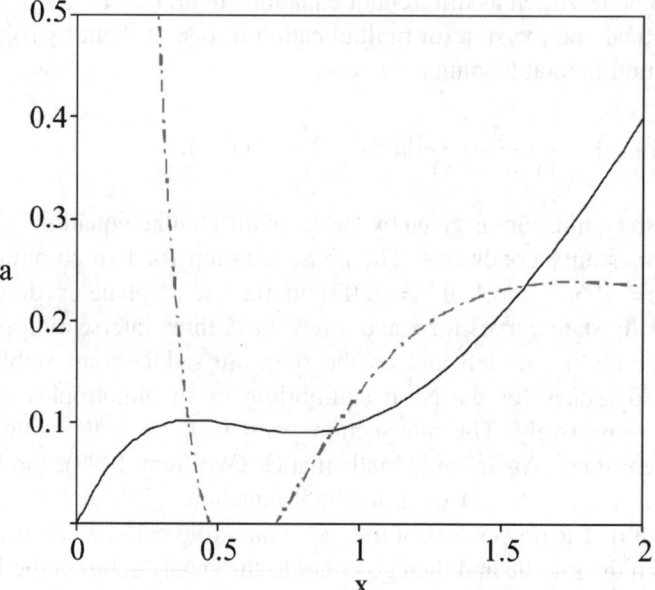

Figure 5a. Phase diagram, Nash equilibrium.

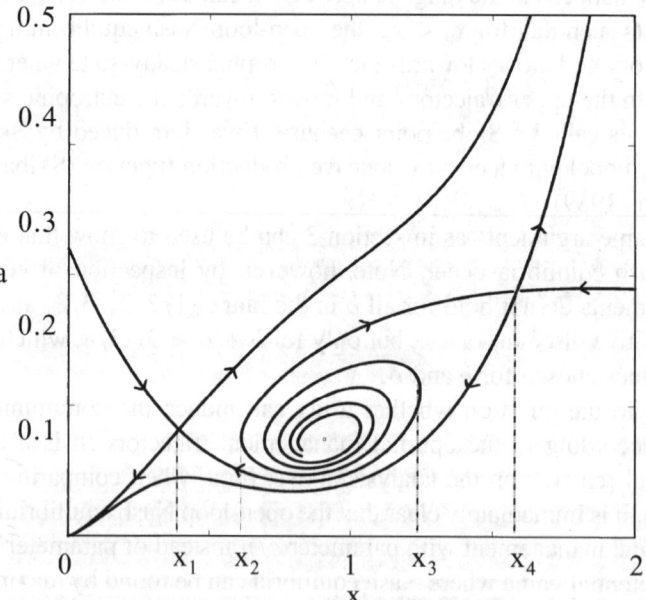

Figure 5b. Stable and unstable manifolds, Nash equilibrium.

but varying number of communities n) can be traced by solving an optimal control problem where the set of differential equations, characterizing the solution, has a parameter c/n. This parameter can be denoted as the bifurcation parameter. It can be shown that in this Hamiltonian system with positive discounting only saddle-node and heteroclinic bifurcations can occur (Wagener 1999; Brock and Starrett 1999). Third, Figure 5b contains similar dynamics as found in a model for external economies with multiple steady-states (Krugman 1991; Matsuyama 1991). Krugman (1991) argues that for initial conditions (history) to the left and to the right of the set of spirals, the economy moves to the left or to the right, respectively, but that for history within the set of spirals, expectations determine where the economy will end up. This is not an answer to uncertainty but since for these initial conditions either a path can be chosen that goes to the left or one that goes to the right, something must determine where the economy will end up. In our model, however, the solution is driven by an objective. A full analysis of the value function shows that the initial conditions determine the outcome, so that only history matters. A Skiba point exists that divides the area of initial conditions into an area that is attracted by the steady-state on the left and an area that is attracted by the steady-state on the right.

4.3. TAXES

Consider the case of achieving the unique steady-state amount of phosphorus under optimal management by a tax τ on phosphorus loading. Under the tax scheme the objectives (7) of the open-loop differential game change to

$$W_i = \int_0^\infty e^{-\rho t}[\ln a_i(t) - \tau(t)a_i(t) - cx^2(t)]dt, i = 1, \ldots, n. \quad (14)$$

The maximum principle requires for the optimal choice of phosphorus loading at each point in time that

$$\frac{1}{a_i(t)} - \tau(t) + \lambda_i(t) = 0, i = 1, \ldots, n. \quad (15)$$

In order to obtain the loading that corresponds to optimal management, it is immediately clear by comparing (8) to (15) that the tax on loading should be chosen such that $\tau(t) = -\lambda(t) + \lambda_i(t)$. This implies that the tax bridges the gap between the social shadow cost of the accumulated phosphorus $\lambda(t)$ and the private shadow cost of the accumulated phosphorus $\lambda_i(t)$ that causes the steady-state phosphorus levels in the open-loop Nash equilibrium to exceed the (unique) steady-state phosphorus level under optimal management. The tax rate, however, is time-dependent, since it has to equalize cooperative and non-cooperative loading at every point in time. Although optimal, such a tax will be very difficult to implement, since it would require a regulating institution to continuously change the tax rate. Another, more realistic, approach would be to choose a fixed tax rate on loading, defined such that the non-cooperative steady-state phosphorus level under the constant tax equals the steady-state phosphorus level under optimal management. This tax will be called the optimal steady-state tax (OSST).

By comparing (9) to (16) and using (10), it is easy to see that the OSST τ^* is given by

$$\tau^* = -\frac{(n-1)\lambda^*}{n} = \frac{(n-1)}{a^*}, \tag{16}$$

where λ^* is the value of the co-state and a^* is total loading in the steady-state under optimal management. Under this constant tax scheme, the open-loop Nash equilibrium will be given by the set of differential equations (2) and, instead of (13),

$$\dot{a}(t) = -[(b+\rho) - \frac{2x(t)}{(x^2(t)+1)^2}][a(t) - \frac{\tau^*}{n}a^2(t)] + 2\frac{c}{n}x(t)a^2(t) \tag{17}$$

with a transversality condition.

It is easy to check that the steady-state (x^*, a^*) for the set of differential equations (2) and (10) under optimal management is also a steady-state for the set of differential equations (2) and (17) in the open-loop Nash equilibrium under the constant tax τ^*. By substituting $a(t) = a^*$ and $\tau^* = (n-1)/a^*$ in the second term between brackets of the right-hand side of equation (17), this term reduces to a^*/n. It is then easy to see that (x^*, a^*) is also a point on the curve representing the steady-states for total loading a in the open-loop Nash equilibrium under the OSST. However, the rest of this curve differs from the one under optimal management.

It should be made clear that the OSST leads to the optimal management steady-state but the path under the OSST that determines the transition to the steady-state *is not the same* as the optimal management path. Coincidence of the optimal management path and the regulated path requires to use the time-dependent tax. To put it differently, the stable manifold of the optimal management problem is not the same as the stable manifold of the regulated problem, although both approach the same saddle-point. Note also that the time of convergence to the steady-state under the OSST will be different than under a time-dependent tax or another control scheme.

If the number of communities $n = 2$, the phase-diagram under the OSST in the (x, a)-plane is drawn in Figure 6a. Although this figure differs from figure 4a for optimal management, it is qualitatively the same. It has one saddle-point and a corresponding stable manifold. Starting at both unregulated Nash equilibrium steady-states, the two communities will change their loadings under the OSST and the equilibrium path will follow this stable manifold and move towards the optimal management steady-state. Starting at the oligotrophic Nash equilibrium, this is a short trajectory, but starting at the eutrophic Nash equilibrium, the path has to bend around the flip-point.

THE ECONOMICS OF SHALLOW LAKES   119

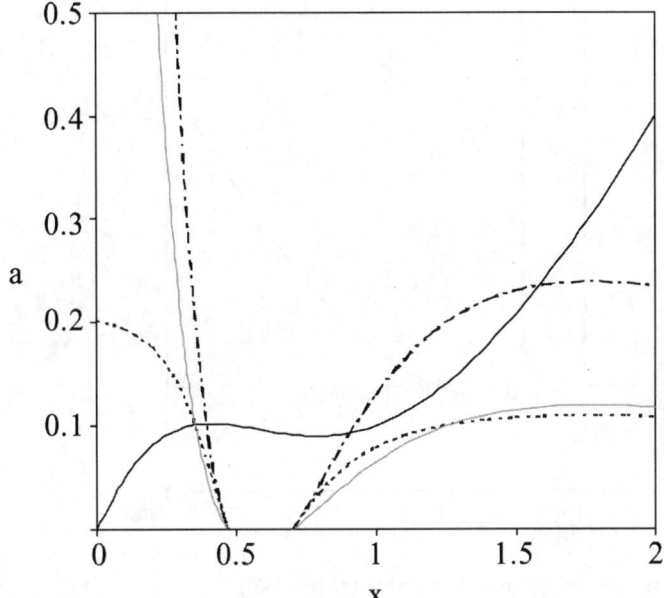

Figure 6a. Phase diagram.
(—— optimal management; – · – · – Nash equilibrium; · · · · · optimal steady-state tax (n = 2))

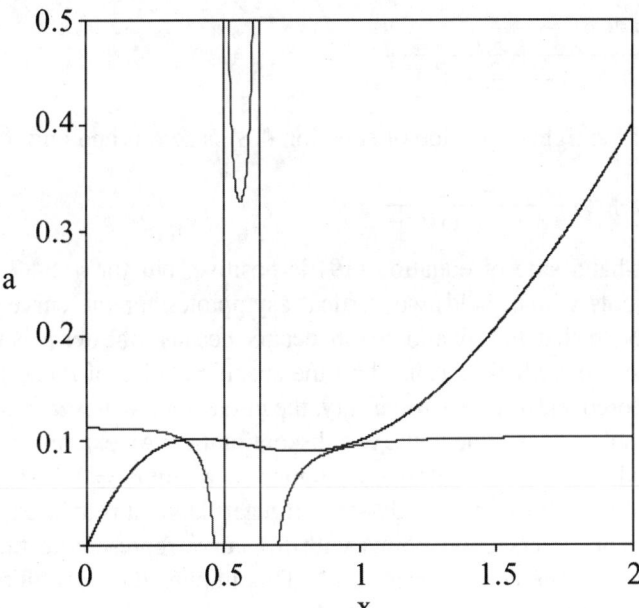

Figure 6b. Phase diagram, optimal steady-state tax (n = 10).

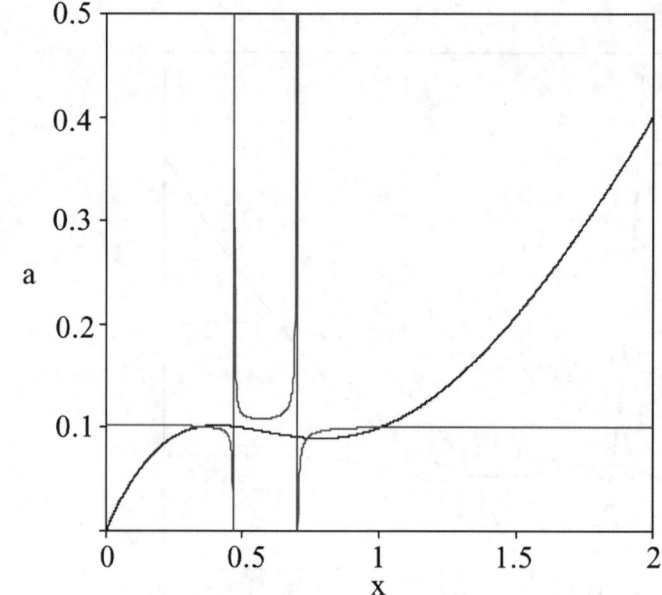

Figure 6c. Phase diagram, optimal steady-state tax (n = 100).

Increasing the number of communities n, at a certain point the phase-diagram under the OSST becomes very complicated. From equation (17), the curve representing the steady-states for a is given by

$$a = \frac{(b+\rho) - \frac{2x}{(x^2+1)^2}}{2\frac{c}{n}x + [(b+\rho) - \frac{2x}{(x^2+1)^2}]\frac{(n-1)}{na^*}}. \tag{18}$$

The denominator of the right-hand side of equation (18) is zero if and only if

$$\frac{2cx}{(n-1)} + [(b+\rho) - \frac{2x}{(x^2+1)^2}]\frac{1}{a^*} = 0. \tag{19}$$

For $n = 2$ the left-hand side of equation (19) is positive, but for $n > 7$ equation (19) has two roots which yield two vertical asymptotes for the curve given by equation (18). Note that this phenomenon occurs because the term between brackets is partly negative which is caused by the specific choice of b and ρ (see also section 4.2). Moreover, if n goes to infinity, the curve approaches $a = a^*$, but this convergence is not uniform, due to the two discontinuities. An example of such a phase-diagram under the OSST is drawn in Figure 6b where $n = 10$. This case still has only one saddle-point stable steady-state. If n gets large, it is to be expected that the curve has more intersection points with the curve representing the lake steady-states, because of the convergence to a^*. This implies the possibility that multiple steady-states occur under the OSST. An example is drawn in Figure 6c where $n = 100$. This case has three steady-states again, two of which are saddle-point stable, whereas the middle one is unstable. For a lower discount rate ρ and a higher number of communities n, it may happen that two more steady-states occur between the asymptotes, one unstable and one saddle-point. The existence of a second steady-state characterized by saddle-point stability leads to the conclusion that if the number of communities n is high, the optimal steady-state tax may not work. Depending on the initial conditions, the OSST may direct the equilibrium path towards a steady-state with a higher phosphorus level than in the optimal management steady-state.

4.4. FEEDBACK NASH EQUILIBRIUM

The open-loop Nash equilibrium is weakly time-consistent but not strongly time-consistent which implies that the equilibrium is not robust against unexpected changes in the state of the lake (Başar 1989). To obtain an equilibrium with the Markov perfect property, the feedback Nash equilibrium has to be found which means that the Hamilton-Jacobi-Bellman equation for the game has to be solved. This is a difficult problem because it is not clear what the value function will look like due to the complexity of the lake model.

In a linear-quadratic framework with quadratic value functions, the solution would be analytically tractable. In a problem like the one under consideration, one would expect that the steady-state amount of phosphorus and total loadings will be higher in the feedback Nash equilibrium than in the open-loop Nash equilibrium. This would confirm the intuition derived in similar type of problems (van der Ploeg and de Zeeuw 1992). If a community knows that the other communities will respond to a higher amount of phosphorus in the lake with lower loadings, it loads more at the margin because loading will be partly offset by the reactions of the other communities. Since all communities argue in this way, total loading in the feedback equilibrium is higher than in case the loadings are not conditioned on the state of the lake as in the open-loop Nash equilibrium. However, Tsutsui and Mino (1990) have shown, for a dynamic duopoly model with sticky prices, that non-quadratic value functions exist that solve the Hamilton-Jacobi-Bellman equation. From this it follows that multiple feedback Nash equilibria exist. The resulting set of steady-state prices lies near to the price under full cooperation or collusion. It implies that feedback Nash equilibria exist with steady-state prices that are better for the duopoly than the steady-state price in the open-loop Nash equilibrium. This technique was applied to the international pollution control model by Dockner and Long (1993) to show that (non-linear) feedback equilibria exist that yield lower steady-state stocks of pollution than the open-loop equilibrium. Since the characteristics of those problems (infinite horizon and a one-dimensional state) are the same as for the lake problem, one may expect that the same technique can be applied here. In this section a few steps will be taken but a full analysis is left for further research.

Suppose that the loading strategies in the (symmetric) feedback Nash equilibrium are given by $a_i = h(x), i = 1, \ldots, n$. The Hamilton-Jacobi-Bellman equation (or dynamic programming equation) for community $i, i = 1, \ldots, n$, becomes

$$\rho V(x) = \max[\ln a_i - cx^2 + V_x(x)[a_i + (n-1)h(x) - bx + \frac{x^2}{x^2+1}]], \quad (20)$$

where V denotes the value function, that is the same for each community.

The first-order condition yields

$$\frac{1}{a_i} = -V_x(x) \rightarrow a_i := h(x) = \frac{-1}{V_x(x)}. \quad (21)$$

Substitution of (21) into (20) leads to

$$\rho V(x) = \ln h(x) - cx^2 - \frac{1}{h(x)}[nh(x) - bx + \frac{x^2}{x^2+1}], \quad (22)$$

and differentiation of (22) with respect to x, using (21) again, leads to

$$-\frac{\rho}{h(x)} = \frac{h'(x)}{h(x)} - 2cx + \frac{h'(x)}{h^2(x)}[-bx + \frac{x^2}{x^2+1}] + \frac{1}{h(x)}[b - \frac{2x}{(x^2+1)^2}]. \quad (23)$$

Rewriting (23) yields an ordinary differential equation in the feedback loading $h(x)$:

$$h'(x)[h(x) - bx + \frac{x^2}{x^2+1}] + h(x)[(\rho + b) - 2cxh(x) - \frac{2x}{(x^2+1)^2}] = 0. \quad (24)$$

If the resulting steady-state x^F were known, equation (2) with total loading $a = nh(x)$ gives a boundary condition for the differential equation (24):

$$h(x^F) = \frac{1}{n}[bx^F - \frac{x^{F2}}{x^{F2}+1}], \quad (25)$$

so that the differential equation can be solved. However, this steady-state is a degree of freedom. This means that it is to be expected that multiple feedback Nash equilibria exist. If that steady-state is chosen to be equal to the steady-state under optimal management, it may yield a feedback Nash equilibrium that sustains the optimal management outcome in steady-state. Note that this does not imply that the same trajectory to the steady-state is followed: welfare will generally still be different. However, it is to be expected that a feedback Nash equilibrium exists that is better than the open-loop one. How can this result be reconciled with the intuition described above and the conclusion that the feedback Nash equilibrium is worse than the open-loop one? Note that the analysis above was restricted to quadratic value functions in a linear-quadratic framework so that linear controls result. It seems that by enlarging the strategy spaces to non-linear controls (with non-quadratic value functions), equilibria arise that are better. This may be recognized as a type of folk-theorem in differential games. Further research is needed to be able to give full answers to these issues.

5. Conclusion

Economics of ecological systems is a much neglected area in the literature. Furthermore, the complex dynamics of these systems and the common property aspect of the ecological services as resource and waste sink, present interesting challenges to economic theory. This paper focuses on the shallow lake, as an example but also because much is known about shallow lakes in the ecological literature. However, the analysis in this paper applies to all models that are driven by convex-concave relations, and models with this feature are very typical for mathematical models in ecology (see Murray 1989).

Internal loading of phosphorus in shallow lakes causes the lake model to be non-linear, with hysteresis effects in the more interesting cases. As a consequence, even if optimal management of the lake has only one steady-state with saddle-point stability, either an increase in the discount rate or an increase in the number of communities, sharing the lake, leads to more saddle-points and complicated dynamics in between. However, a Skiba point exists, which means that in these cases the initial level of accumulated phosphorus determines whether the lake will end up in a clear or a turbid state. For a small number of communities, a constant tax on the loading of phosphorus can induce optimal behaviour and a return to a clear state, but for a large number this policy may not work. The analysis employs the open-loop Nash equilibrium to characterize non-cooperative behaviour. The feedback Nash equilibrium would be more appropriate but is very difficult to identify for these type of problems. Some first steps are given but a full analysis is left for further research.

References

Başar, T. and G. J. Olsder (1982), *Dynamic Noncooperative Game Theory*. New York: Academic Press.
Başar, T. (1989), 'Time Consistency and Robustness on Equilibria in Noncooperative Dynamic Games', in F. van der Ploeg and A. J. de Zeeuw, eds., *Dynamic Policy Games in Economics* (pp. 9–54). North Holland, Amsterdam: Contributions to Economic Analysis 181.
Brock, W. A., S. R. Carpenter and D. Ludwig (1997), *Notes on Optimal Management of a Lake Subject to Flips*, Working Paper. University of Wisconsin.
Brock, W. A. and D. Starrett (1999), *Nonconvexities in Ecological Management Problems*, Working Paper. University of Wisconsin.
Brock, W. A. and A. G. Malliaris (1989), *Differential Equations, Stability and Chaos in Dynamic Economics*. Amsterdam: North Holland.
Brock, W. A. and A. de Zeeuw (2002), 'The Repeated Lake Game', *Economics Letters* **76**, 109–114.
Carpenter, S. R. and K. L. Cottingham (1997), 'Resilience and Restoration of Lakes', *Conservation Ecology* **1**, 2.
Carpenter, S. R., D. Ludwig and W. A. Brock (1999), 'Management of Eutrophication for Lakes Subject to Potentially Irreversible Change', *Ecological Applications* **9**(3), 751–771.
Dechert, W. D. and W. A. Brock (1999), *Lakegame*, Working Paper. University of Houston.
Dockner, E. J. and N. V. Long (1993), 'International Pollution Control: Cooperative Versus Noncooperative Strategies', *Journal of Environmental Economics and Management* **24**, 13–29.
Krugman, P. (1991), 'History Versus Expectations', *Quarterly Journal of Economics* **CVI**, 651–667.

Ludwig, D., D. D. Jones and C. S. Holling (1978), 'Qualitative Analysis of Insect Outbreak Systems: The Spruce Budworm and Forest', *Journal of Animal Ecology* **47**, 315–332.

Ludwig, D., B. H. Walker and C. S. Holling (1997), 'Sustainability, Stability and Resilience', *Conservation Ecology* **1**, 7.

Matsuyama, K. (1991), 'Increasing Returns, Industrialization, and Indeterminacy of Equilibrium,' *Quarterly Journal of Economics* **CVI**, 617–650.

Monderer, D. and L. S. Shapley (1996), 'Potential Games', *Games and Economic Behaviour* **14**, 124–143.

Murray, J. D. (1989), *Mathematical Biology*. Berlin: Springer Verlag.

Scheffer, M. (1997), *Ecology of Shallow Lakes*. New York: Chapman and Hall.

Skiba, A. K. (1978), 'Optimal Growth with a Convex-concave Production Function', *Econometrica* **46**(3), 527–539.

Tsutsui, S. and K. Mino (1990), 'Nonlinear Strategies in Dynamic Duopolistic Competition with Sticky Prices', *Journal of Economic Theory* **52**, 136–161.

Van der Ploeg, F. and A. J. de Zeeuw (1992), 'International Aspects of Pollution Control', *Environmental & Resource Economics* **2**, 117–139.

Wagener, F. O. O. (1999), *Shallow Lakes*, Working Paper. University of Amsterdam.

Wolfram, S. (1999), *The Mathematica Book*, 4th ed. Cambridge: Wolfram Media/Cambridge University Press.

Appendix A: The Skiba Point

As also noted later in the main text, the open-loop Nash equilibrium results from maximizing the welfare objective (a potential function)

$$W = \int_0^\infty e^{-\rho t} [\sum_{i=1}^n \ln a_i(t) - cx^2(t)] dt. \tag{A1}$$

What follows is strongly based on Wagener (1999).

Define the Hamiltonian function

$$H = e^{-\rho t} g(x, a_1, \ldots, a_n) + \mu f(x, a_1, \ldots, a_n), \tag{A2}$$

where

$$g(x, a_1, \ldots, a_n) = \sum_{i=1}^n \ln a_i - cx^2, \ f(x, a_1, \ldots, a_n) = \sum_{i=1}^n a_i - bx + \frac{x^2}{x^2 + 1}, \tag{A3}$$

and define the current value Hamiltonian function

$$\tilde{H} = g(x, a_1, \ldots, a_n) + \lambda f(x, a_1, \ldots, a_n), \lambda = e^{\rho t} \mu. \tag{A4}$$

The maximum principle yields the necessary conditions (8), (9) and (2) with the parameter c replaced by c/n, which then yields the set of differential equations (2) and (13) for the open-loop Nash equilibrium with the phase-diagram given in Figure 5a and the stable and unstable manifolds in Figure 5b. Denote the x-coordinate of the oligotrophic steady-state as x_1 and of the eutrophic steady-state as x_4, and denote the range with curls as $[x_2, x_3]$, where x_2 and x_3 are the x-coordinates of the intersection points of the outer upper curl and the outer lower curl with the curve representing the steady-states for $x(f = 0)$, respectively.

Along trajectories, it holds that

$$\frac{dH}{dt} = \frac{\partial H}{\partial x}\frac{dx}{dt} + \frac{\partial H}{\partial \mu}\frac{d\mu}{dt} + \frac{\partial H}{\partial t} = \frac{\partial H}{\partial x}\frac{\partial H}{\partial \mu} - \frac{\partial H}{\partial \mu}\frac{\partial H}{\partial x} + \frac{\partial H}{\partial t} = -\rho e^{-\rho t}g. \quad (A5)$$

It follows that

$$\tilde{H}(0) = H(0) = -\int_0^\infty \frac{dH}{dt}dt = \int_0^\infty \rho e^{-\rho t} g\, dt = \rho W. \quad (A6)$$

Furthermore,

$$\frac{dW}{dx} = \frac{dW}{dt}\frac{dt}{dx} = \frac{1}{\rho}\frac{d\tilde{H}}{dt}\frac{1}{f} = \frac{1}{\rho}[\frac{\partial \tilde{H}}{\partial x}f + f(\rho\lambda - \frac{\partial \tilde{H}}{\partial x})]\frac{1}{f} = \lambda. \quad (A7)$$

Condition (8) yields

$$\lambda = -\frac{1}{a_i}, i = 1, \ldots, n, \to \lambda = -\frac{n}{a}(a = \sum_{i=1}^{n} a_i). \quad (A8)$$

The proof of the existence of a unique Skiba point takes four steps.

1) Suppose the initial condition x_0 is in the interior of the range $[x_2, x_3]$.

It is better to jump immediately to the upper trajectory instead of to the *same* trajectory some point earlier, or to the lower trajectory instead of to the *same* trajectory some point earlier, because the welfare difference

$$\int \frac{dW}{dx}dx = \int \lambda dx = \int -\frac{n}{a}dx < 0. \quad (A9)$$

2) Suppose the initial condition is x_2. The choice is either to jump to the upper trajectory and start at the intersection point with $f = 0$ or to jump to the lower trajectory, by a proper choice of intial loadings a. Because

$$\frac{\partial W}{\partial a} = \frac{1}{\rho}\frac{\partial \tilde{H}}{\partial a} = \frac{1}{\rho}\sum_{i=1}^{n}\frac{\partial}{\partial a_i}(g + \lambda f) = \frac{1}{\rho}\sum_{i=1}^{n}\frac{1}{a_i^2}f = \frac{1}{\rho}\frac{n^3}{a^2}f \quad (A10)$$

and $f < 0$ below the intersection point, the welfare difference between the upper trajectory and the lower trajectory is negative, so that it is better to jump to the lower trajectory at x_2.

3) Suppose the initial condition is x_3. The choice is either to jump to the lower trajectory and start at the intersection point with $f = 0$ or to jump to the upper trajectory. Because above the intersection point $f > 0$, it follows from equation (A10) that the welfare difference between the upper trajectory and the lower trajectory is positive, so that it is better to jump to the upper trajectory at x_3.

4) Compare now the upper trajectory leading to steady-state on the right, with co-state λ_2, and the lower trajectory leading to the steady-state on the left, with co-state λ_1. Denote the welfare difference as ΔW. Using the results in steps 2–3 and equations (A7)–(A8), it follows that

$$\Delta W(x_2) < 0, \Delta W(x_3) > 0, \frac{d}{dx}\Delta W = \lambda_2 - \lambda_1 > 0. \quad (A11)$$

From (A11) it follows that a point x_S in the interior of the range $[x_2, x_3]$ exists, such that

$$\Delta W(x_S) = 0; \ \Delta W(x) < 0, x \in [x_2, x_S); \ \Delta W(x) > 0, x \in (x_S, x_3]. \tag{A12}$$

The point x_S is called a Skiba point and (A12) implies that if the initial amount of phosphorus x_0 is on the left-hand side of the Skiba point, the equilibrium jumps to the lower trajectory and moves towards the steady-state on the left and if the initial amount of phosphorus x_0 is on the right-hand side of the Skiba point, the equilibrium jumps to the upper trajectory and moves towards the steady-state on the right.

Multiple Species Boreal Forests – What Faustmann Missed*

ANNE-SOPHIE CRÉPIN
The Beijer International Institute of Ecological Economics, The Royal Swedish Academy of Sciences, Box 50005, 10405 Stockholm, Sweden (E-mail: asc@beijer.kva.se)

Abstract. Recent research in natural sciences shows that the dynamics in boreal forests are much more complex than what many models traditionally used in forestry economics reflect. This essay analyzes some challenges of accounting for such complexity. It shows that the optimal harvesting strategy for forest owners is history dependent and for some states of the forest, more than one strategy may be optimal. This paper confirms earlier literature on shallow lakes and coral reefs and shows that this kind of phenomena seem much more common than previously thought. They are valid for a wide range of ecosystems that cover large surfaces and they do not depend on the choice of some specific function to model the non-linearity. There are also indications that theses results could be obtained even for resources with concave growth if at least one species with non-linear growth affects their dynamics.

Key words: forestry, multiple steady states, non-linear, Skiba points

1. Introduction

Research inspired by Holling (1973) and May (1977) showed that ecosystem dynamics were complex: not accounting for them would lead to serious surprises. Non-linearity, interactions between species, disturbances, and threshold effects were some examples of patterns that play a crucial role in ecosystems' dynamics. What does this imply for resource management?

Theories for economic management of renewable natural resources have recently experienced drastic changes due to the model for shallow lakes that Scheffer (1998) and Carpenter and Cottingham (1997) produced independently of one other. This model accounts for the possibility that such lakes may flip between a clear and a turbid state. Brock and Starrett (2003) gave a complete treatment of the problem of optimal management of shallow lakes. Mäler, Xepapadeas and de Zeeuw (2003) provided a dynamic economic analysis of shallow lakes managed under common property and developed a method to solve such a differential game. Wagener (2003) gave a local criterion that ensured the existence of *Skiba points* (Skiba 1978) – initial states with more than one optimal path – when discount rates were small.

---
* This paper is a shorter version of Chapter 2 in Crépin (2002).

All these contributions are based on the study of shallow lakes and could for that reason be criticized for their narrow applicability. After all, shallow lakes are not a dominant ecosystem on our planet. Crépin (2002, Ch. 1) showed that similar results – existence of multiple interior steady states and Skiba points – could be obtained for coral reef fisheries. So many results from fisheries literature (Gordon 1954; Scott 1955; Clark 1973; Clark 1976) do not necessarily hold in the presence of threshold effects like in coral reefs. Coral reefs cover a small area on our planet but they play a key role as a nursery for fish living in the open seas (Hughes et al. 2003). So these results may have larger consequences for human beings than previously thought.

Another criticism against the shallow lake literature, which applies to the coral reef article as well, is that non-linearity is modeled in the same way in all the articles. Each of them uses a specific sigmoid function to model threshold effects. If we are unlucky, the results obtained in the shallow lake and coral reef literature could be due to this function's special features, which may have little in common with real phenomena.

This article takes up both these criticisms. It shows that the results from the shallow lake literature can be obtained even for boreal forests – terrestrial ecosystems, also called *taiga*, that occupy a wide belt around the arctic circle in the northern hemisphere. It also shows that these results do not depend on the use of the specific sigmoid function.

Extensive literature exists on ecosystem modeling and management where several species interact. In particular, Pastor and others described the complexity of boreal forest ecosystems (Pastor and Mladenoff 1992; Pastor et al. 1996; Danell et al. 1998). They pointed out the importance of the interaction of species and non-linearities. Pastor et al. grasped the most important dynamics in boreal forests by using a system of three differential equations that represent the three-species' dynamics. This paper uses a slightly modified version of Pastor et al.'s model. The model is used to analyze the challenges that ecosystem complexity implies when calculating management rules for forestry.

This paper derives some optimal management rules to guide forest owners or decision makers who harvest several species and also value the standing forest for recreational purposes. The harvesting rules are continuous, which means that the decision maker calculates how much it is optimal to harvest at each point in time. The boreal forests are considered as any other ecosystem that produces renewable resources. So a similar approach as in the fisheries literature (Clark 1976) can be used to find out optimal harvesting paths. This is not usual in the forestry literature. Models of forestry management – initiated by Faustmann (1849), Pressler (1860), and Ohlin (1921) – typically assume that it is optimal to harvest the whole stand at discrete time intervals. These models result in rules that determine when the whole stand of one specific species should be harvested. Bergland, Ready and Romstad (forthcoming) made recently an interesting contribution to this literature. They studied optimal management of moose and pine in Norway and calculated an optimal rotation period and optimal moose harvest in a model in which there was feedback between moose and pine. Crépin (2002, Ch. 2) determined the specific conditions required for whole stand harvesting to be optimal and calculated optimal harvesting rules in the special case studied here. The optimal rotation period has typically different length, which depends on the presence of other species in the forest. The non-convexity of the growth function for pine implies that several rotation periods may be optimal.

Section 2 discusses modelling issues, presents the ecosystem model of a boreal forest, and analyzes its dynamic properties. Section 3 models different harvesting regimes. Section 4 presents the computer simulations. Section 5 presents the conclusion.

2. A Three-species Boreal-forest Model

A good ecosystem model should grasp the most important dynamics in the real ecosystem and still be simple enough to give useful understanding. What characterizes boreal forests? Compared to many other ecosystems (coral reefs, rainforests), most boreal forests host few species. A small amount of these species appear in large populations and dominate the forests. Boreal tree species are coniferous trees that include spruce and pine and some hardwood species such as aspen and birch. They give the forests their special look and influence their dynamics. Large browsers (moose, elk) can influence boreal forests' dynamics because they feed on the trees and can damage them. Boreal forests also host various types of lower vegetation, which vary depending on humidity and nutrients levels in the soil (Bernes 1994; Danell et al. 1998; Pastor et al. 1997). The model used here describes only the dynamics of conifers, caduceus trees and large browsers. Similarly to Pastor et al. (1997), the influence of other species on the forests' dynamics is neglected. This paper uses two models for a boreal forest: a very general model and a specific model, which is useful for computer simulations.

2.1. GENERAL MODEL

Murray (1993) and Gurney and Nisbet (1998) presented methods to build and analyze ecosystem models. The very general three-species model (SYS) could represent the dynamics of boreal forests:

$$\dot{x} = G_x(x, y, z)$$
$$\dot{y} = G_y(x, y, z) \quad \text{(SYS)}$$
$$\dot{z} = G_z(x, y, z)$$

where x, y, and z are vectors that represent categories of browsers (moose), caduceus trees (birch), and conifers' (pine) biomasses, respectively. $G_i(x, y, z)$ is the natural growth of species or category $i \in \{x, y, z\}$ in the absence of harvesting. Such a general model can embed almost any kind of specific dynamics among three species groups. This is useful to model age or space, for example. Negative biomass values are impossible and impose the restrictions:

$$x \geq 0, y \geq 0 \text{ and } z \geq 0 \tag{1}$$

This is equivalent to imposing that $G_i(x, y, z)$ must be non-negative when species i is extinct ($i = 0$). Assume x_0, y_0, and z_0 are the initial stock of species at time $t = 0$. If none of the restrictions (1) are binding, and for every $i \in \{x, y, z\}$, $G_i(x, y, z)$ and its derivatives regarding x, y, and z are continuous, then SYS has a unique solution:[1]

$$x(t) = \varphi_x(x_0, y_0, z_0, t) \tag{2}$$
$$y(t) = \varphi_y(x_0, y_0, z_0, t)$$
$$z(t) = \varphi_z(x_0, y_0, z_0, t)$$

In section 3, the SYS model is used to calculate general forestry management rules. Computer simulations require a more precise specification of forests' dynamics. Ideally, one would like to represent forest ecosystems as accurately as possible, including all relevant species, their ages, and spatial distribution. Some simplifications are necessary because such a representation would be rather difficult to work with. This paper focuses on the effects of species' interactions so the model used does not account for age and space.

2.2. SPECIFIC MODEL

John Pastor et al. (1997) presented a simplified boreal-forest model in which conifers, caduceus trees, and herbivores interacted. A slightly modified version is used here, where pine, birch, and moose are the three interacting species. In Pastor's model, species can move in space. This is not assumed here, and the model represents one homogenous piece of land. Removing the spatial dimension from Pastor's model implies that conifers grow in an uncontrolled way, which is not very realistic. Still, conifers tend to come rather late in the succession of species and are invasive (Bernes 1994). A convex-concave growth function for conifers is one way to represent such a feature.

For $i, j \in \{x, y, z\}$, let r_i be species i's growth rates and a_{ij} be interaction coefficients of species j on i. Then SYS can be rewritten with growth functions defined by (3). This specific model is called SYS1.

$$G_x(x, y, z) = x - x^2 + a_{xy}xy + a_{xz}xz \tag{3}$$
$$G_y(x, y, z) = r_y y - y^2 - a_{yx}xy - a_{yz}z$$
$$G_z(x, y, z) = r_z z^2 - z^3 - a_{zx}xz - a_{zy}y$$

In SYS1 $r_x = a_{xx} = a_{yy} = a_{zz} = 1$. Crépin (2002, Ch. 2) showed that models with general values of these parameters had dynamic properties similar to SYS1's after parameter calibration. So studying SYS1 produces necessary information on more general models.

Moose feed on birch and pine, so both tree species have a positive effect on moose biomass, which is proportional to tree stocks ($a_{xy}xy$ and $a_{xz}xz$). The corresponding effect of moose on tree biomass is negative ($-a_{yx}xy$ and $-a_{zx}xz$). The negative term $-x^2$ describes the crowding effect that occurs when the moose population becomes too large. The negative term $-y^2$ describes that birch is shade intolerant. When the density is too high, birch stop multiplying because new plants cannot get enough light. These terms are in quadratic form because in the absence of any other species moose and birch would have logistic growth by assumption and $\frac{1}{a_{ii}} = 1, i \in \{x, y\}$ would then be the carrying capacity[2] for species i.

In contrast, pine exhibits a convex-concave growth, which is modeled using the function $z^2(r_z - z)$. When pine biomass is small, young pines establish better with increasing biomass, so growth is convex because the term $r_z z^2$ dominates. When the population becomes larger, competition arises and growth becomes concave because the term $-z^3$ dominates. When carrying capacity ($\frac{1}{a_{zz}} = 1$) is reached, more pine leads to negative growth. Any function of the form $z^q(r_z - z)$ has similar dynamic properties. So why specify $q = 2$? First, this is convenient because we avoid large exponents or an extra parameter q. Second, q influences the size of the domain on which of the growth function for z is convex. Large values of q imply a larger domain in which pine growth is convex. This means that if we choose $q = 2$, we depart just a little from the traditional concavity assumption. So if the traditional results do not hold for $q = 2$, they are likely to hold even less for larger values of q. The terms $a_{yz}z$ and $a_{zy}y$ represent the effects of competition between species of trees.

Figure 1 shows what a species' growth looks like when species interaction is not accounted for. Note that $G_y(x, 0, z)$ and $G_z(x, y, 0)$ are negative so the constraints (1) may be binding for some initial points. Crépin (2002, Ch. 2) describes what happens when constraints are binding.

While a detailed analysis of SYS1's dynamics can also be found in Crépin (2002, Ch. 2), the important features are summarized here. Depending on parameter values, SYS1 has up to 15 feasible steady states. Their characteristics are in Table I.[3]

Figure 2 show a three-dimensional phase diagram of SYS1. Table II summarizes the directions of motions in the different regions separated by the manifolds $P_{x1}(\dot{x} = 0, x \neq 0)$, $P_y(\dot{y} = 0)$, and $P_z(\dot{z} = 0)$.

The phase space is also divided into regions so that trajectories that start anywhere in one region will end up in the same stable equilibrium. The manifolds called *separatrices* (Kuznetsov 1995) separate these different regions. The separatrices are difficult to locate in the three-dimensional phase space, but Figure 3 shows the phase diagram when moose, for example, have disappeared. The dotted curved line shows the separatrices' approximate location in this special case.

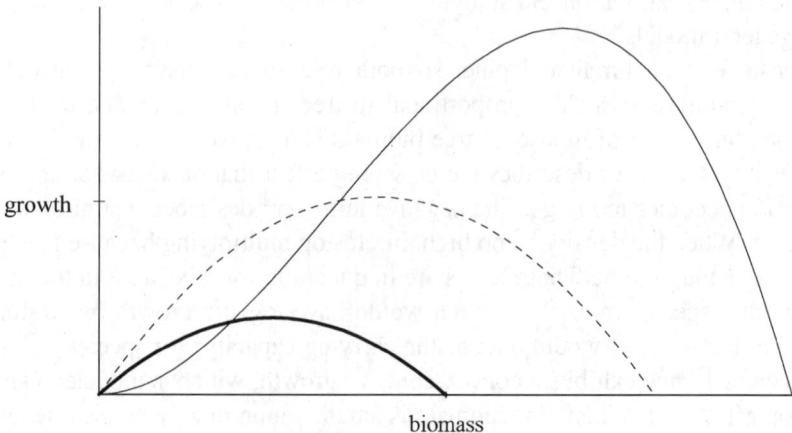

Figure 1. Growth patterns for moose (bold line), birch (dashed line), and pine (thin line).

Table I. Steady states properties

Ecosystem	Steady state	Dynamic properties
extinction	$S0 = (0, 0, 0)$	one unstable or saddle
one species	$S1_x = (1, 0, 0)$	one stable or saddle
	$S1_y = (0, r_y, 0)$	one stable or saddle
	$S1_z = (0, 0, r_z)$	one stable
two species	$S2_\kappa = \left(0, \kappa^2 \frac{r_z - \kappa}{a_{zy}}, \kappa\right)$	one stable, two saddle, one unstable
	$S2_\phi = (1 + a_{xz}\phi, 0, \phi)$	one stable, one saddle
	$S2_g = \left(\frac{1+a_{xy}r_y}{a_{xy}a_{yx}+1}, \frac{r_y - a_{yx}}{a_{xy}a_{yx}+1}, 0\right)$	one stable
three species	$S3_\kappa = (X(\kappa), Y(\kappa), \kappa)$	one stable, three saddle

Table II. Directions of motion in different regions

Region	$\dot{x}$	$\dot{y}$	$\dot{z}$
I: below P_{x1}, P_y and P_z	> 0	> 0	> 0
II: below P_{x1}, P_y above P_z	> 0	> 0	< 0
III: below P_{x1}, P_{z1} above P_y	> 0	< 0	> 0
IV: below P_z, P_y above P_{x1}	< 0	> 0	> 0
V: below P_z above P_{x1} and P_y	< 0	< 0	> 0
VI: below P_{x1} above P_z and P_y	> 0	< 0	< 0
VII: below P_y above P_z and P_y	< 0	> 0	< 0
VIII: above P_{x1}, P_y and P_z	< 0	< 0	< 0

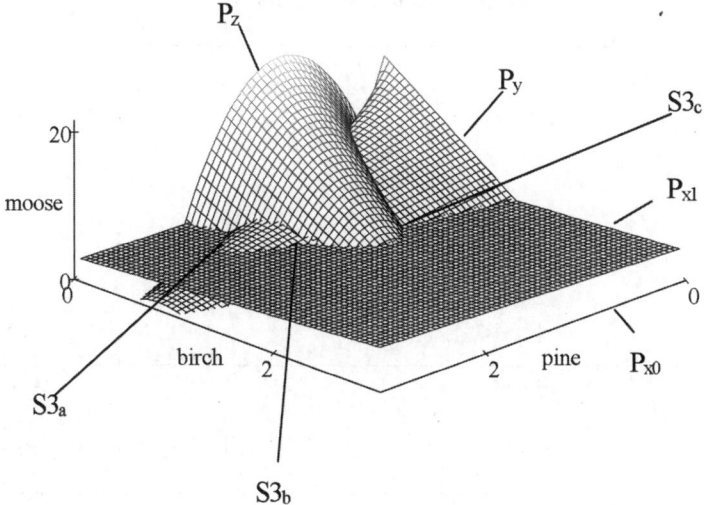

Figure 2. Phase diagram.

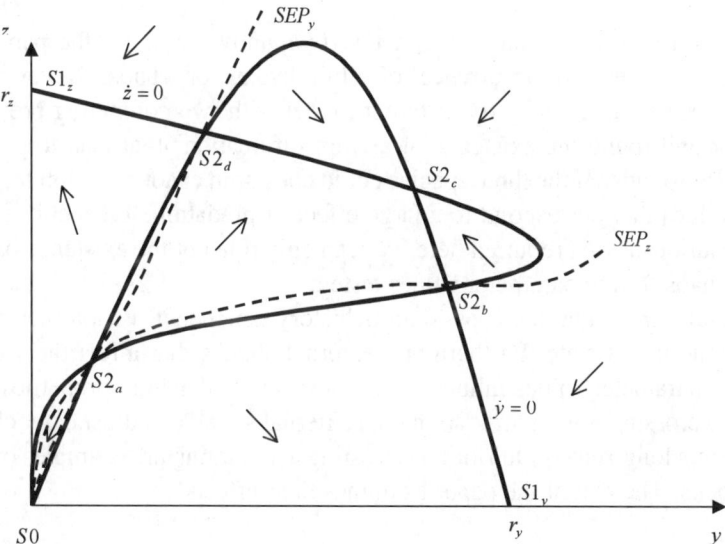

Figure 3. Phase diagram when x = 0.

Steady states' simulations with sets of arbitrarily chosen parameters show the fold bifurcations[4] that might occur. For example, when the birch growth rate varies, six fold bifurcations can be distinguished for values of r_y around 0.5, 0.65, 1.2, 1.45, 1.84, and 1.86. Figure 4 represents pine biomass in a steady state for different birch growth-rate values. It shows clearly the fold bifurcations that occur. For low growth rates, no feasible interior steady is stable. Birch cannot maintain itself and becomes extinct. Similar results are obtained when the other parameters vary.

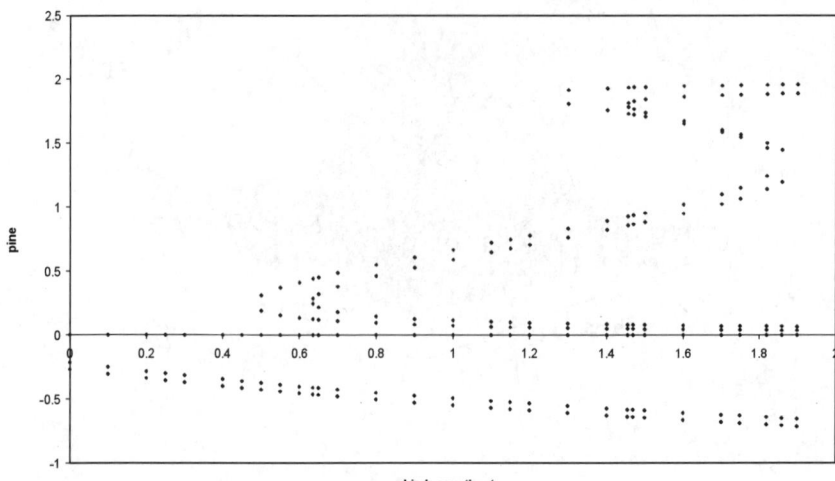

Figure 4. Fold bifurcations (pine).

Computation of the Lyapunov exponents (Lyapunov 1892) for the parameter values tested showed no occurrence of limit cycles or chaos. Nevertheless, Takeuchi (1996) studied a slightly different model with two competing preys and one predator and found the existence of a Hopf bifurcation[5] that lead to periodic orbits. He also found that the three species could coexist in chaotic motion for some parameter values that correspond to a large effect of predation. Takeuchi's results prove that cautiousness is required here. We can only rule out the existence of limit cycles and chaos for the parameter values tested.

These results show that the ecosystem is history dependent: its long-term state depends on the initial state. Furthermore, external shocks that affect the variable stocks or the parameter values influence the ecosystem's dynamic properties. This could lead to crossing a separatrix so that an external shock can drastically change the ecosystem's long-run equilibrium. Harvesting and hunting are examples of such external shocks. The rest of the paper examines their effects.

3. General Management Rules

What happens when harvesting is introduced in ecosystem SYS? Let $h = (h_i)_{i \in \{x,y,z\}}$ be a vector of harvests at time t. To begin with, assume that this vector is arbitrary. The SYS system is transformed into:

$$\dot{x} = G_x(x, y, z) - h_x \qquad (4)$$
$$\dot{y} = G_y(x, y, z) - h_y$$
$$\dot{z} = G_z(x, y, z) - h_z$$
$$x \geq 0, y \geq 0, z \geq 0$$

It is easy to verify that harvesting affects the separatrices' location and thereby the stable states' basins of attraction. Harvesting may also cause bifurcations; the number of equilibria and their dynamic properties can then differ from the unexploited ecosystem case.

Assume the initial stock of species at time $t = 0$ is x_0, y_0, and z_0. If none of the species becomes extinct, then (4) has a unique solution:

$$x(t) = \Phi_x(x_0, y_0, z_0, t) \quad (5)$$
$$y(t) = \Phi_y(x_0, y_0, z_0, t)$$
$$z(t) = \Phi_z(x_0, y_0, z_0, t)$$

Given that owners harvest forest species to increase their welfare, the next step is to decide what harvesting rules maximize welfare in each period.

Assume forest owners accounted for the benefits that they can continuously extract from all of the forest's species. To maximize the forest's net benefits, they wanted to find out how much of each species to harvest every time. These benefits were harvesting profits and the forest's net environmental and recreational values. Modeling recreational and environmental values requires that species are entered as state variables directly into the objective function that is maximized.

Let $\Omega_i(h_i)$ represent profits from harvesting species i and $\Omega_e(x, y, z)$ represent net benefits from environmental and recreational services. Functions Ω_i and Ω_e were assumed to be concave. Assume further that the owners gave different positive weights K_x, K_y, K_z and K_e to respective net forest benefits. At time t, the total net benefits were $B(h, x, y, z) = K_x\Omega_x(h_x) + K_y\Omega_y(h_y) + K_z\Omega_z(h_z) + K_e\Omega_e(x, y, z)$. Owners faced the problem:

$$\max_h \int_0^{+\infty} B(h, x, y, z)e^{-\rho t} dt \quad (6)$$
$$\text{s.t. } \dot{x} = G_x(x, y, z) - h_x$$
$$\dot{y} = G_y(x, y, z) - h_y$$
$$\dot{z} = G_z(x, y, z) - h_z$$
$$x \geq 0, y \geq 0 \text{ and } z \geq 0$$

Pontryagin et al. (1964) developed methods to solve such optimal control problems. This paper uses the method found in Arrow and Kurz (1970, Ch. 2) to solve such problems when there are non-negativity constraints on state variables. Hestenes (1966) and Seierstad and Sydsæter (1987) have developed similar methods.

Let $h^*(t) = (h_i^*(t))_{i \in \{x,y,z\}}$ represent harvest choices, which are admissible solutions for problem (6). If the constraint qualification (Kuhn and Tucker 1951) was true then there existed functions of time, $\lambda_i(t)$ so that for each t, $\mathcal{H}$ is the current-value Hamiltonian and $\mathcal{L}$ is the Lagrange function:

$$\mathcal{H}(x,y,z,h,\lambda,t) = \sum_i (K_i\Omega_i(h_i) + \lambda_i(G_i(x,y,z) - h_i)) \qquad (7)$$
$$+ K_e\Omega_e(x,y,z)$$
$$\mathcal{L}(x,y,z,h,\lambda,\mu,t) = \mathcal{H}(x,y,z,h,\lambda,t) + \sum_i \mu_i(G_i(x,y,z) - h_i) \qquad (8)$$

Then $h^*(t)$ maximizes $\mathcal{H}(x,y,z,h,\lambda,t)$ subject to the constraints $G_i(x,y,z) - h_i \geq 0$ for all $i \in \{x,y,z\}$, for which $i(t) = 0$. Further $\dot{\lambda}_i = \rho\lambda_i - \frac{\partial \mathcal{L}}{\partial i}$, evaluated at $i = i(t)$, $h = h^*(t)$, $\lambda = \lambda(t)$. The Lagrange multipliers μ_i must be such that for all i, $\frac{\partial \mathcal{L}}{\partial h_i} = 0$ for $(x,y,z) = (x(t),y(t),z(t))$, $h = h^*(t)$, $\lambda = \lambda(t)$, and $\mu_i(t)i(t) = 0$, $\mu_i(t)(G_i(x,y,z) - h_i) = 0$. The necessary conditions for $h^*(t)$ to be optimal amount to:

1) the equations of motion for the exploited ecosystem,
$$\dot{x} = G_x(x,y,z) - h_x \qquad (9)$$
$$\dot{y} = G_y(x,y,z) - h_y$$
$$\dot{z} = G_z(x,y,z) - h_z$$

2) the necessary conditions for optimal harvest, $\forall i \in \{x,y,z\}$,
$$K_i \frac{\partial \Omega_i(h_i^*)}{\partial h_i} - \lambda_i - \mu_i = 0 \text{ or } h_i^* = 0 \qquad (10)$$

3) the shadow price equations for each species, $\forall i, j \in \{x,y,z\}$,
$$\dot{\lambda}_j = \rho\lambda_j - K_e \frac{\partial \Omega_e(x,y,z)}{\partial j} - \sum_{i \in \{x,y,z\}} (\lambda_i + \mu_i) \frac{\partial G_i(x,y,z)}{\partial j} \qquad (11)$$

4) the conditions for the multipliers' non-negativity,
$$\forall i \in \{x,y,z\}, \mu_i(t)\dot{i}(t) \geq 0, \mu_i(t)(G_i(x,y,z) - h_i) = 0 \qquad (12)$$

Conditions (10) are sufficient for optimal harvests because the profits from harvests are additively separable and concave in each harvests, which implies that $\mathcal{H}$ is concave in harvests. Proposition 1 follows directly from equation (10):

Proposition 1
The optimal size for each species' harvest is such that the marginal value from harvesting more of the species equals the marginal value of retaining more of it in the ecosystem.

Assume that the equation system (10) has a solution: for $i \in \{x,y,z\}$, the solution is unique because the profit functions are concave and it has the form $h^* = \psi(\lambda)$. So the optimal harvests h_i^* are:
$$h_i^* = \psi_i(\lambda) \text{ or } h_i^* = 0 \qquad (13)$$

Conditions (11) imply that in a steady state, for all $j \in \{x, y, z\}$,

$$\rho \lambda_j = K_e \frac{\partial \Omega_e(x, y, z)}{\partial j} + \sum_{i \in \{x,y,z\}} (\lambda_i + \mu_i) \frac{\partial G_i(x, y, z)}{\partial j}$$

Proposition 2
In a steady state, the interest on a species' marginal value in the ecosystem equals the species' marginal environmental benefit plus the species' marginal benefit in maintaining its own and other species' stock.

Together, propositions 1 and 2 imply that environmental benefits and other species' stocks must affect harvest size. Whether or not the harvest is higher or lower depends on the species' effects on its own and other species' growth rate, and on the environmental benefits.

If $\widehat{\mathcal{H}}(x, y, z, \lambda, t) \equiv \max_h \mathcal{H}(x, y, z, h, \lambda, t)$, is a concave function of (x, y, z) for given λ and t, then any policy is optimal that satisfies the conditions (9)–(12) and the transversality conditions (14). Arrow and Kurz (1970) provide these sufficiency conditions for infinite horizon problems.

$$\lim_{t \to +\infty} e^{-\rho t} \lambda_i(t) \geq 0, \quad \lim_{t \to +\infty} e^{-\rho t} \lambda_i(t) i(t) = 0 \tag{14}$$

Let $\Delta_r(x, y, z)$ be the principal minor of order r in the Hessian for $\widehat{\mathcal{H}}(x, y, z, \lambda, t)$. The maximized Hamiltonian $\widehat{\mathcal{H}}$ is concave on $\mathbb{R}^3_+$ if and only if for all points (x, y, z) and for all Δ_r, $(-1)^r \Delta_r(x, y, z) \geq 0$ for $r = \{1, 2, 3\}$. In case this condition does not hold, the sufficiency conditions for optimum are not satisfied, and there may be either several, one or no optimal solutions. If there is more than one solution candidate, the comparison of benefits between different solution paths may be necessary to determine, which one is optimal.

When the forest is optimally exploited, two cases can be distinguished and must be analyzed separately.
- Case 1: no species ever becomes extinct.
- Case 2: at least one of the species becomes extinct at some point t_1 in time. This case is a bit complicated because the analysis differs depending on which species disappears first. Crépin (2002, Ch. 2) analyzed this case when SYS1 was optimized.

If no species became extinct, the exploited system would follow the equations of motion given by (15):

$$\begin{aligned}
\dot{x} &= G_x(x, y, z) - \psi_x(\lambda_x) \\
\dot{y} &= G_y(x, y, z) - \psi_y(\lambda_y) \\
\dot{z} &= G_z(x, y, z) - \psi_z(\lambda_z) \\
\dot{\lambda}_x &= \rho \lambda_x - K_e \frac{\partial \Omega_e(x, y, z)}{\partial x} - \sum_{i \in \{x,y,z\}} \lambda_i \frac{\partial G_i(x, y, z)}{\partial x}
\end{aligned} \tag{15}$$

$$\dot{\lambda}_y = \rho\lambda_y - K_e \frac{\partial \Omega_e(x,y,z)}{\partial y} - \sum_{i \in \{x,y,z\}} \lambda_i \frac{\partial G_i(x,y,z)}{\partial y}$$

$$\dot{\lambda}_z = \rho\lambda_z - K_e \frac{\partial \Omega_e(x,y,z)}{\partial z} - \sum_{i \in \{x,y,z\}} \lambda_i \frac{\partial G_i(x,y,z)}{\partial z}$$

Brock and Malliaris (1989) showed methods to study such dynamic systems. Assume the initial species stocks were x_0, y_0, and z_0 at $t = 0$ and the initial shadow prices λ_{x_0}, λ_{y_0}, and λ_{z_0}. Then if the system (15)'s right hand side satisfied the Lipschitz condition (see Brock and Malliaris 1989), it had a unique solution defined for every species $i \in \{x, y, z\}$ by:[6]

$$i^*(t) = \Phi_i(x_0, y_0, z_0, \lambda_{x_0}, \lambda_{y_0}, \lambda_{z_0}, t)$$

$$\lambda_i^*(t) = \Phi_{\lambda_i}(x_0, y_0, z_0, \lambda_{x_0}, \lambda_{y_0}, \lambda_{z_0}, t)$$

Assume there is an optimal solution and that the system (15) had at least one steady state.[7] Crépin (2002, Appendix) then showed that the eigenvalues of such steady state came in pairs α, $\rho - \alpha$. So the saddle-point properties proven in Kurz (1968) remain, even when several steady states existed and the Hamiltonian was not concave. Proposition 3 follows directly:

Proposition 3
Suppose $\rho > 0$ then, in the neighborhood of a steady state; the system (15) is either totally unstable or has the instability characterized by the saddle-point property.

If the system has several steady states, then for each of them, there is a limit value for ρ, say $\tilde{\rho}$, under which the steady state exhibits a local saddle-path property or has eigenvalues equal to zero and above which the steady state is locally unstable. This produces a series $(\tilde{\rho})$ of threshold values for ρ. Corollary 4 follows directly.

Corollary 4
Suppose the system (15) has several steady states. Let $\underline{\rho} = \min(\tilde{\rho})$ and $\overline{\rho} = \max(\tilde{\rho})$. If $\rho < \underline{\rho}$, there is a local saddle path that lead toward each steady state. If $\rho > \overline{\rho}$, all of the steady states are locally unstable. If $\underline{\rho} > \rho > \overline{\rho}$, some steady states are locally unstable while others have a local saddle path.

Note that often in such problems $\underline{\rho} = 0$ and for $0 < \rho < \overline{\rho}$, there is an odd number of steady states, which come in consecutive pairs of saddle points and unstable states (Birkhoff 1927).

The occurrence of several steady states implies that there is no obvious optimal trajectory from a given starting point. Candidate trajectories toward different equilibria must be compared to determine, which one is optimal. Relation (16) follows from the classic Hamilton-Jacobi result and makes welfare comparisons possible when there is more than one candidate-optimal steady state.

$$\int_0^{+\infty} B(h^*, x, y, z)e^{-\rho t}dt = \frac{1}{\rho}\mathcal{H}(x_0, y_0, z_0, h_0^*, \lambda_0) \tag{16}$$

This relation can also be used to localize Skiba points, which are initial states with more than one optimal path (Beyn et al. 2001). Note that as Deissenberg et al. (2001) reminded us, if strict concavity is not given, Skiba points generically do not coincide with the unstable steady states and the latter are not necessarily optimal. Wagener (2003) showed that for systems with one state and one control, if there was a cusp bifurcation (see Kuznetsov 1995 for a definition) when the discount rate was zero, then for small positive discount rates, the system had a Skiba point. This result has not yet been generalized to higher dimensions.

Section 4 simulates a specific model of a boreal forest with continuous, multiple-use harvesting.

4. Simulations with a Specific Model

The simulations aim to answer the following questions: How many steady states exist and are they feasible? What is the optimal path in the exploited forest? What are the dynamics at each steady state? What do the dynamics of the system look like when they are not in a steady state? What is the basin of attraction of the steady states where no species become extinct? What happens in boundary points? Do bifurcations occur when parameters vary?

Kuznetsov (1995), Judd (1998), and Beyn et al. (2001) showed several useful numerical methods. MATHCAD[8] simulated steady states and the dynamics in their neighborhood.[9] Steady state analysis is not enough to obtain a quality picture of the dynamics of the systems of differential equations that have been studied. Each system may have very complex dynamics outside of the steady states. Limit cycles[10] or chaotic attractors[11] may very well be present. Systems' simulations using DYNAMICS[12] (Nusse and Yorke 1997) help picture the systems' dynamics outside steady states. In particular, this program can explore the system for limit cycles and chaos. MATLAB[13] solved two point boundary value problems and localized Skiba points in the optimized system using the method developed in Beyn et al. (2001).

4.1. MULTIPLE-USE SPECIFIC MODEL

Let q_i and c_i be positive constants that represent the unit price and some cost parameters, respectively, for species $i \in \{x, y, z\}$. Then the profit from harvesting species i can be expressed as $\Omega_i(h_i) = q_i h_i - c_i h_i$. Let $\alpha_i \equiv K_i c_i$ and $p_i \equiv \frac{q_i}{c_i}$; then the weighted profit from harvesting species i is $K_i \Omega_i(h_i) = \alpha_i h_i(p_i - h_i)$. This function has a convex cost part $(\alpha_i h_i^2)$ and is concave in harvest.[14]

Let ζ, η, and θ be constant weights associated with species x, y, and z respectively. $\zeta, \eta, \theta \in (0, 1)$ and $\zeta + \eta + \theta \leq 1$. Environmental and recreational benefits from the forest are then $\Omega_e(x, y, z) = x^\zeta y^\eta z^\theta$. Ω_e is increasing and concave in all species's stocks and the environmental benefits from the forest are zero as soon as one species becomes extinct. This is a strong assumption but the functional form still accounts for important characteristics of environmental and recreational benefits: it increases with number of species and populations sizes.[15]

The multiple-use problem from section 3 is now transformed into:

$$\max_h \int_0^{+\infty} \left(\sum_i \alpha_i h_i (p_i - h_i) + K_e x^\zeta y^\eta z^\theta \right) e^{-\rho t} dt$$

s.t. $\dot{x} = x - x^2 + a_{xy}xy + a_{xz}xz - h_x$
$\dot{y} = r_y y - y^2 - a_{yx}xy - a_{yz}z - h_y$
$\dot{z} = r_z z^2 - z^3 - a_{zx}xz - a_{zy}y - h_z$

The Lagrange function is still given by (8), and the Hamiltonian (7) is rewritten with appropriate functional forms:

$$\mathcal{H}(x, y, z, h, \lambda, t) = \sum_i \alpha_i h_i (p_i - h_i) + K_e x^\zeta y^\eta z^\theta$$
$$+ \lambda_x (x - x^2 + a_{xy}xy + a_{xz}xz - h_x)$$
$$+ \lambda_y (r_y y - y^2 - a_{yx}xy - a_{yz}z - h_y)$$
$$+ \lambda_z (r_z z^2 - z^3 - a_{zx}xz - a_{zy}y - h_z)$$

The concavity conditions (17) are complicated. They are computed in appendix 5 for $K_e = 0$:

$$\lambda_x \geq 0 \qquad (17)$$
$$4\lambda_x \lambda_y \geq (\lambda_x a_{xy} - \lambda_y a_{yx})^2$$
$$\lambda_z (6z - 2r_z) \left(4\lambda_x \lambda_y - (\lambda_x a_{xy} - \lambda_y a_{yx})^2 \right) \geq 2\lambda_y (\lambda_x a_{xz} - \lambda_z a_{zx})^2$$

They imply that the marginal value of having moose and birch in the forest, respectively, must be positive. The marginal value of having pine in the forest must be positive if z^* is relatively large ($z^* > \frac{r_z}{3}$), and negative if z^* is relatively small ($z^* < \frac{r_z}{3}$). In each case, z^* must be large enough or small enough to satisfy the third sufficiency condition. When $\frac{r_z}{3}$, the pine population has reached one-third of its carrying capacity. For smaller pine populations, the pines' growth rate is convex. If the pine population is larger, it is concave (recall Figure 1).

For $K_e > 0$, that is when recreational and environmental benefits enter the social welfare function, concavity conditions are very tedious to compute. Similarly to when $K_e = 0$, concavity conditions are not always satisfied, which implies that welfare comparisons between trajectories are necessary and the maximizing problem may have one, several or no optimal solutions, depending on the initial state. As pointed out in section 3, comparing the value of the Hamiltonian for different initial states can help sort out different optimal trajectory candidates.

Table III. Parameter values in the main simulation

parameter	r_y	r_z	a_{xy}	a_{xz}	a_{yx}	a_{yz}	a_{zx}	a_{zy}	ρ	α
value	1.6	2	1	0.5	0.1	0.2	0.05	0.7	0.02	1
parameter	β	γ	K_e	q_x	q_y	q_z	ζ	η	θ	
value	1	1	0.1	1	2	2	1/3	1/3	1/3	

The optimal harvest is written, for all i:

$$h_i^* = \frac{1}{2}\left(p_i - \frac{\lambda_i + \mu_i}{\alpha_i}\right), \text{ or } h_i^* = 0$$

When no species are depleted by (12), the Lagrange multipliers must equal zero at every point in time ($\mu_i = 0$). If shadow prices are large enough compared to costs ($\forall i, p_i > \frac{\lambda_i}{\alpha_i}$), all harvests are positive. Assuming that this is the case, an optimal trajectory $(x^*, y^*, z^*, \lambda^*)$ must solve SYS2.

$$\dot{x} = x - x^2 + a_{xy}xy + a_{xz}xz - \frac{1}{2}p_x + \frac{\lambda_x}{2\alpha_x} \quad \text{(SYS2)}$$
$$\dot{y} = r_y y - y^2 - a_{yx}xy - a_{yz}z - \frac{1}{2}p_y + \frac{\lambda_y}{2\alpha_y}$$
$$\dot{z} = r_z z^2 - z^3 - a_{zx}xz - a_{zy}y - \frac{1}{2}p_z + \frac{\lambda_z}{2\alpha_z}$$
$$\dot{\lambda}_x = -K_e \zeta x^{\zeta-1} y^\eta z^\theta - \lambda_x(1 - 2x + a_{xy}y + a_{xz}z - \rho) + \lambda_y a_{yx} y + \lambda_z a_{zx} z$$
$$\dot{\lambda}_y = -K_e \eta x^\zeta y^{\eta-1} z^\theta - \lambda_x a_{xy} x - \lambda_y(r_y - 2y - a_{yx}x - \rho) + \lambda_z a_{zy}$$
$$\dot{\lambda}_z = -K_e \theta x^\zeta y^\eta z^{\theta-1} - \lambda_x a_{xz} x + \lambda_y a_{yz} - \lambda_z(2r_z z - 3z^2 - a_{zx}x - \rho)$$

4.2. SIMULATIONS

Giving numerical values to parameters helps simulate this differential equation system. Table III shows the values that were assessed to the parameters in the main simulation.

Note that there are some reasons to treat cases $K_e = 0$ and $K_e > 0$ separately. First the system's dynamics are much easier to simulate when $K_e = 0$. Second $K_e = 0$ can also be interpreted as the case of a private owner who only maximizes profits from forestry. Then $\alpha_i = c_i$ for all $i \in \{x, y, z\}$: it is reasonable to believe that private owners do not care about where profits come from, because they have no redistribution goals.

Even when parameters are replaced with numeric values, SYS2's steady states cannot be analytically computed. They must be evaluated numerically, which

Table IV. Steady states

Steady state	x	y	z	λ_x	λ_y	λ_z
$S3_a^*$	**1.767**	**0.503**	**1.138**	**−0.078**	**1.53**	**0.672**
$S3_b^*$	0.979	0.367	0.971	−0.71	1.555	0.668
$S3_c^*$	**0.234**	**0.624**	**1.207**	**0.067**	**1.294**	**0.591**
$S3_d^*$	0.537	0.231	0.036	0.235	1.407	2.32
$S2_a^*$	2.14	2.837	0	−6.263	10.233	2
$S2_b^*$	2.263	1.525	0	−0.186	2.46	2

implies that the number of steady states could be underestimated. Usually, numerically evaluated steady states depend on quality initial guesses; different initial guesses can lead to different steady states. For each set of tested parameter values, the steady states were simulated with MATHCAD using a thousand different randomly chosen initial values between zero and ten for each variable.[16]

For the benchmark's parameter values, with $K_e = 0$, SYS2 has four feasible interior steady states ($S3_a^*$, $S3_b^*$, $S3_c^*$, $S3_d^*$) and one steady state with a negative pine biomass, which is not feasible. Three of them ($S3_a^*$, $S3_c^*$ and the infeasible state) are saddle points and the other two are unstable with only two eigenvalues with negative real part. This is not surprising given the result stated in proposition 3 and its corollary 4. The existence of a saddle steady state, with negative pine biomass, indicates that for some initial states, depletion of all pine may be optimal. This possibility was studied in Crépin (2002, Ch. 2).[17] The SYS2 system was simulated to check what happened when pine disappeared. This produced two steady states with no pine. Table IV shows the variables' values in all steady states. The steady states in bold are saddle points.

So for the benchmark's parameter values with $K_e = 0$, the optimized forest characterized by SYS2 had at least six feasible steady states, of which at least two lead to pine depletion. It is easy to verify that steady state $S3_c^*$ is the only equilibrium that satisfies the sufficiency conditions (17).

Simulations[18] using the method developed in Beyn et al. (2001) showed the existence of Skiba points. For example such points were $(4.98, 0.5, 1.1)$, $(2, 1.33, 1.1)$, and $(2, 0.5, 1.86)$. In those points, the welfare obtained by going to steady state $S3_a^*$ was about the same as the welfare obtained when going to $S3_c^*$. This shows that $S3_a^*$ could sometimes be optimal even though the sufficiency conditions do not hold. Whether or not this is true depends on initial conditions.

The sets of all Skiba points form the Skiba manifolds of this system. Locating all Skiba points would be very tedious. According to Beyn et al., one could use the initial state variables as continuation parameters to approximate the Skiba manifold. Depending on whether or not pine extinction can ever be optimal there might be Skiba manifolds that separate the regions with optimal interior states from the regions where a steady state with no pine is optimal.

The effects of variations in the birch growth rate on respective species' biomasses in steady states were simulated in the case when environmental benefits were not accounted for ($K_e = 0$). These simulations also showed bifurcations for low birch growth rates. For r_y below 0.45, there was only one feasible steady state; for r_y above 0.65, there were four and in between, there were three steady states. Variations in other parameters' value also lead to bifurcations.

When a social planner accounted for environmental benefits, the results were modified. Some simulations showed that the number of steady states was generally higher for relatively low birch growth rates and smaller for higher birch growth rates. Comparing the steady states in both management regimes shows that the saddle points have higher levels of each population when environmental benefits are accounted for. This is not true for the unstable equilibria, which are probably not optimal but this is not proved yet. For the benchmark's parameter values, there are only three steady states and no infeasible saddle point. This hints that the risk of pine becoming extinct is much lower when environmental benefits are accounted for, which is not surprising.

5. Concluding Remarks

This paper shows that a boreal forest, managed to maximize values derived from it, may have several optimal interior steady states. Each optimal state is reachable by following an optimal saddle path. On this path, the marginal value derived from harvesting each species equals the marginal value of retaining the species in the ecosystem at each point in time. The path that is optimal to follow and thus the optimal steady state depends on the initial state of the system. The paper shows also that for some initial states – Skiba points – there are more than one optimal trajectory, which lead to different steady states.

The possibility of multiple steady states was already pointed out in Clark (1990) but this was assumed to occur only in specific problems of resource economics. This paper reinforces results from earlier literature on shallow lakes and coral reefs and shows thereby that this kind of phenomena is probably much more common than previously thought. The non-concave pine growth is an essential assumption that determines the results. This paper shows that:
- The results obtained in the shallow lake and coral reef literature cited in introduction are not due to specific properties of the sigmoid function used in these articles. Rather, they depend on the non-linearity present in the equation system. It does not seem to matter how this non-linearity is modeled.
- Non-linearity in pine growth affects growth and populations of moose and spruce and their harvest. So the results may be obtained even if there was non-linearity in the dynamics of a species that is not even harvested, as long as the species affects the harvested species. This is a crucial point because

if this is a general finding and if many ecosystems host species with non-linear dynamics that affect harvested species' dynamics, this result could be generalized to most natural resources produced within ecosystems. In that case, traditional models of natural resource management derived from Gordon (1954), Scott (1955), Clark (1973) and Clark (1976) would almost always fail in the long-run.

The existence of Skiba points means that at those points there are multiple accounting prices for the ecosystem resources. This may have important implications for valuation studies: which price should be chosen as accounting price in a Skiba point?

What do these results imply for a manager? The existence of multiple steady states reveals the exploited ecosystem's dependency on history. What is optimal for one state of the world is not necessarily optimal for another. This has implications for the management of such systems. If the system had had a unique interior steady state, the forest owner could have reached the optimal saddle path by harvesting just enough for marginal costs to equal marginal benefits from harvest – including costs and benefits due to environmental changes and effects on other species. Here, margin analysis is usually not enough to determine the optimal trajectory at given initial points. One needs more information. Some of the steady states are minimizing rather than maximizing solutions and can easily be ruled out. For the remaining maximizing steady state solutions, the optimal solution depends on the initial state.

Furthermore if an optimal trajectory passes close to a Skiba manifold, small management mistakes could lead to the manifold's crossing. If this happens another trajectory becomes optimal and the future harvest opportunities can be completely modified. Exogenous changes in the system can also lead to a Skiba manifold crossing. Such exogenous changes include any changes that affect the variables, such as diseases, storms, and exogenous market shocks.

A forest owner searching to maximize the values obtained from the forest would need to know for sure all future harvesting benefits and costs to find out which trajectory is optimal. Forest owners are usually not so knowledgeable. In fact costs and benefits become more and more uncertain the further the time horizon is. So one cannot once and for all determine the optimal management tactics.

What should forest owners do? For example, they may want to revise their tactics at more or less regular intervals to see whether they need to adapt to changes or not. This is difficult because at each revision time they need to gather new information about all future costs and benefits, which is costly. This cost and the potential benefits obtained from adapting to changes give forest owners indications about how often they should recalculate the management tactics. The managers could also try to detect the thresholds between the basins of attraction of the different steady states. This means that they need to identify the Skiba manifolds or at least find a way to approximate their location.

Acknowledgement

Special thanks to Karl-Göran Mäler for many discussions about the paper. I also thank Kjell Danell, Partha Dasgupta, Karl-Gustaf Löfgren, Eric Nævdal, John Pastor, Hans Wallin, Angels Xabadia, Göran Ågren, an anonymous referee and participants in a seminar given at the department of forestry economics at the Swedish University of Agricultural Sciences in Umeå, January 2002. Thanks to the participants of the Meeting of the Theory group of the Resilience Network, May 3–5, 2000 in Belize (especially William Brock and Simon Levin), who gave me valuable comments on an early draft. I gratefully acknowledge funding from the Swedish Council for Forestry and Agricultural Research (SJFR) and the Swedish Research Council for Environment, Agricultural Sciences and Spatial Planning (Formas).

Notes

1. See Kuznetsov (1995) for a proof.
2. Carrying capacity refers to the population size that can be supported in the area studied. It is the population size for which natural growth is equal to zero when there are no interactions from other species.
3. In Table I, κ solves $a_{yz}a_{zy}^2 - r_y r_z a_{zy} Z + r_y a_{zy} Z^2 + Z^3 r_z^2 - 2r_z Z^4 + Z^5$, which has at most four real roots. $\phi = \frac{1}{2}(r_z - a_{zx}a_{xz} \pm \sqrt{((r_z - a_{zx}a_{xz})^2 - 4a_{zx})})$. $X(\kappa) = \frac{a_{zy}+a_{xz}a_{zy}\kappa+a_{xy}r_z\kappa^2-a_{xy}\kappa^3}{a_{zx}a_{xy}\kappa+a_{zy}}$, $Y(\kappa) = \kappa \frac{-a_{zx}+(r_z-a_{zx}a_{xz})\kappa-\kappa^2}{a_{zy}+a_{xz}a_{zx}\kappa}$.
4. A fold bifurcation is a bifurcation that corresponds to the presence of an eigenvalue equal to zero. When this happens, two equilibria *collide* and disappear (Kuznetsov 1995).
5. A Hopf bifurcation corresponds to the presence of complex conjugate eigenvalues with zero real parts.
6. This is the case when $\forall i \in \{x, y, z\}$, $\psi_i(\lambda_i)$ and its derivatives with regard to λ_i are continuous.
7. Unfortunately, the usual existence theorems cannot be applied to guarantee the existence of an optimal solution because the concavity conditions have not necessarily been met. The existence of a steady state of the system (15) is also not guaranteed.
8. Mathcad is a software used to solve math problem. Both numerical and analytical methods can be used.
9. A copy of this program code can be requested from the author.
10. A limit cycle is an isolated cycle of a continuous time dynamic system. A cycle is a periodic orbit, that is a non-quilibrium orbit such that a trajectory starting at a point will return to the same point after a time period called *period*. See Kuznetsov (1995) for a more detailed definition.
11. An attractor is roughly a subset of the phase space toward which the initial conditions may be attracted. An attractor is said to be chaotic when and if we take two typical points on the attractor that are separated from each other by a small distance; then, for increasing time, these points move apart exponentially fast. Thus a small uncertainty in the initial state of the system rapidly leads to the inability to forecast its future. See Grebogi et al. (1987) for further reading and references on the topic.
12. DYNAMICS is a program that explores the dynamics of differential and difference equation systems.
13. MATLAB is a matrix based interactive program doing numeric computation and data visualization.

14. This relies on these assumptions: recreational benefits from moose hunting are neglected and profits are assumed to be independent from moose density; timber harvesting has no effect on the timber's market price; and there are no returns to scale.
15. To measure diversity, alternatives to the Cobb Douglas function can be found in Stirling and Wilsey (2001) or Norberg et al. (2001).
16. Different ranges and different numbers of initial guesses were also tested for some parameter values and did not produce additional steady states.
17. Note that the general saddle point property stated in proposition 3 does not remain when some species are extinct.
18. The program codes can be requested from the author.

References

Arrow, K. J. and M. Kurz (1970), *Public Investment, the Rate of Return, and Optimal Fiscal Policy*. Baltimore and London: The John Hopkins Press.
Bergland, O., R. Ready and E. Romstad (forthcoming) 'Differentiating at the Speed of Light in a Universe of Trees and Moose', in T. Aronsson, R. Axelsson and R. Brännlund, eds., *Environmental and Resource Economics in Honor of Karl-Gustaf Löfgren*.
Bernes, C. (1994), *Biologisk Mångfald I Sverige – En Landstudie*. Naturvårdsverket.
Beyn, W.-J., T. Pampel and W. Semmler (2001), 'Dynamic Optimization and Skiba Sets in Economics Examples', *Optimal Control Applications and Methods* **22**, 251–280.
Birkhoff, G. D. (1927), *Dynamical Systems*, Vol. 9 of *Colloquium Publications*. USA: American Mathematical Society, 1966 edition.
Brock, W. A. and A. G. Malliaris (1989), *Differential Equations, Stability and Chaos in Dynamic Economics*, Advanced Textbooks in Economics. The Netherlands: North-Holland Elsevier Science Publishers.
Brock, W. A. and D. Starrett (2003), 'Managing Systems with Non-Convex Positive Feedback', *Environmental and Resource Economics* **26**, 575–602.
Carpenter, S. R. and K. L. Cottingham (1997), 'Resilience and Restoration of Lakes', *Conservation Ecology* **1**(1), 1–17. http://www.consecol.org/Journal/vol1/iss1/art2/index.html.
Clark, C. (1976), *Mathematical Bioeconomics: The Optimal Management of Renewable Resources*. New York: Wiley.
Clark, C. W. (1973), 'The Economics of Overexploitation', *Science* **181**, 630–634.
Clark, C. W. (1990), *Mathematical Bioeconomics – The Optimal Management of Renewable Resources*, Pure and Applied Mathematics. John Wiley and Sons Inc, second edition. First published in 1976.
Crépin, A.-S. (2002), 'Tackling the Economics of Ecosystems'. Ph.D. thesis, Stockholm University, Department of Economics. Dissertations in economics 2002: 6.
Danell, K., T. Willebrand and L. Baskin (1998), 'Mammalian Herbivores in the Boreal Forests: Their Numerical Fluctuation and Use by Man', *Conservation Ecology* **2**(2:9), 1–18. http://www.consecol.org/vol2/iss2/art9.
Deissenberg, C., G. Feichtinger, W. Semmler and F. Wirl (2001), *History Dependence and Global Dynamics in Modles with Multiple Equilibria*. Germany: Working Paper No. 12 Center for Empirical Macroeconomics, Department of Economics, University of Bielefeld P.O. Box 100 131, 33501 Bielefeld.
Faustmann, M. (1849), 'Berechnung Des Wertes Welchen Waldboden Sowie Noch Nicht Haubare Holzbestände Für Die Waldwirtschaft Besitzen', *Allgemeine Forst und Jagd-Zeitung* **25**, 441–455.
Gordon, H. S. (1954), 'The Economic Theory of a Common Property Resource: The Fishery', *Journal of Political Economy* **62**, 124–142.

Grebogi, C., E. Ott and J. A. Yorke (1987), 'Chaos, Strange Attractors, and Fractal Basin Boundaries in Nonlinear Dynamics', *Science*, 632–638.
Gurney, W. and R. Nisbet (1998), *Ecological Dynamics*. Oxford University Press.
Hestenes, M. R. (1966), *Calculus of Variations and Optimal Control Theory*. United States: John Wiley and Sons.
Holling, C. S. (1973), 'Resilience and Stability of Ecological Systems', *Annual Review of Ecology and Systematics* **4**, 1–23.
Hughes, T. P., A. H. Baird, D. R. Bellwood, M. Card, S. R. Connolly, C. Folke, R. Grosberg, O. Hoegh-Goldberg, J. B. C. Jackson, J. Kleypas, J. M. Lough, P. Marshall, M. Nyström, S. R. Palumbi, J. M. Pandolf, B. Rosen and J. Roughgarden (2003), 'Climate Change, Human Impacts and the Resilience of Coral Reefs', *Science* **301**, 929–933, August.
Judd, K. L. (1998), *Numerical Methods in Economics*. The MIT Press.
Kuhn, H. and A. Tucker (1951), 'Non-Linear Programming', in J. Neyman, ed., *Proceedings of the Second Berkeley Symposium on Mathematical Statistics and Probability* (pp. 481–492). Berkeley and Los Angeles: University of California Press.
Kurz, M. (1968), 'The General Instability of a Class of Competitive Growth Processes', *The Review of Economic Studies* **35**(102), 155–174.
Kuznetsov, Y. A. (1995), *Elements of Applied Bifurcation Theory*, Applied Mathematical Sciences 112. Springer Verlag.
Lyapunov, A. (1892), *General Problem of Stability of Motion*. Kharkov: Mathematics Society of Kharkov.
Mäler, K.-G., A. Xepapadeas and A. de Zeeuw (2000), 'The Economics of Shallow Lakes', *Environmental and Resource Economics* **26**, 603–624.
May, R. M. (1977), 'Thresholds and Breakpoints in Ecosystems with Multiple Stable States', *Nature* **269**, 471–477.
Murray, J. D. (1993), *Mathematical Biology*, Biomathematics Texts. Springer, second, corrected edition.
Norberg, J., D. P. Swaney, J. Dushoff, J. Lin, R. Caqsagrandi and S. A. Levin (2001), 'Phenotypic Diversity and Ecosystem Functioning in Changing Environments: A Theoretical Framework', *Proceedings of the National Academy of Sciences, USA* **98**(20), 11376–11381.
Nusse, H. E. and J. A. Yorke (1997), *Dynamics: Numerical Explorations*, Vol. 101 of *Applied Mathematical Sciences*. New York, USA: Springer-Verlag, second, revised and enlarged edition. Accompanying Computer Program Dynamics 2 Coauthored by Brian R. Hunt and Eric J. Kostelich.
Ohlin, B. (1921), 'Till Frågan Om Skogarnas Omloppstid', *Ekonomisk Tidskrift* **22**, 89–113.
Pastor, J., B. Dewey, R. Moen, D. J. Mladenoff, M. White and Y. Cohen (1997), 'Spatial Patterns in the Moose-Forest-Soil Ecosystem on the Isle Royale, Michigan', *Ecological Applications* **8**(2), 411–424.
Pastor, J. and D. J. Mladenoff (1992), 'The Southern Boreal-Northern Hardwood Forest Border', in H. H. Shugart, R. Leemans and G. B. Bonan, eds., *A System Analysis of the Global Boreal Forest* (pp. 216–240). Cambridge University Press.
Pastor, J., D. J. Mladenoff, Y. Haila, J. Bryant and S. Payette (1996), 'Biodiversity and Ecosystem Processes in Boreal Regions', in H. A. Mooney, J. H. Cushman, E. Medina, O. E. Sala and E. D. Schulze, eds., *Functional Roles of Biodiversity: A Global Perspective*, Chapt. 3 (pp. 33–69). John Wiley and Sons Ltd.
Pontryagin, L., V. Boltyanskii, R. Gamkrelidze and E. Mishchenko (1964), *The Mathematical Theory of Optimal Processes*. Pergamon Press. Translated by D.E. Brown.
Pressler, M. R. (1860), 'Aus der Holzzuwachlehre (Zweiter Artikel)', *Allgemeine Forst- und Jagdzeitung* **36**, 173–191.
Scheffer, M. (1998), *Ecology of Shallow Lakes*, Vol. 22 of *Population and Community Biology Series*. Chapman and Hall, first edition.

Scott, A. D. (1955), 'The Fishery: The Objectives of Sole-Ownership', *Journal of Political Economy* **63**, 116–124.

Seierstad, A. and K. Sydsæter (1987), *Optimal Control Theory with Economic Applications*, Vol. 24 of *Advanced Textbooks in Economics*. North Holland: Elsevier Science Publisher.

Skiba, A. K. (1978), 'Optimal Growth with a Convex-Concave Production', *Econometrica* **46**(3), 527–539.

Stirling, G. and B. Wilsey (2001), 'Empirical Relationships Between Species Richness, Evenness and Proportional Diversity', *The American Naturalist* **158**(3), 286–299.

Takeuchi, Y. (1996), *Global Dynamical Properties of Lotka-Volterra Systems*. World Scientific Publishing Co. Pte. Ltd.

Wagener, F. (2003), 'Skiba Points and Heteroclinic Bifurcation Points, with Applications to the Shallow Lake System', *Journal of Economic Dynamics and Control* **27**(9), 1533–1561.

Appendix

Let $\Delta_r(x, y, z)$ be the principal minor of order r in the Hessian for $\widehat{\mathcal{H}}(x, y, z, \lambda, t)$. The maximized Hamiltonian $\widehat{\mathcal{H}}$ is concave on R_+^3 if and only if for all points (x, y, z) and for all Δ_r, $(-1)^r \Delta_r(x, y, z) \geq 0$ for $r = 1, 2, 3$. The Hessian for $\widehat{\mathcal{H}}(x, y, z, \lambda, t)$ is given by:

$$\begin{bmatrix} -2\lambda_x & \lambda_x a_{xy} - \lambda_y a_{yx} & \lambda_x a_{xz} - \lambda_z a_{zx} \\ \lambda_x a_{xy} - \lambda_y a_{yx} & -2\lambda_y & 0 \\ \lambda_x a_{xz} - \lambda_z a_{zx} & 0 & \lambda_z(2r_z - 6z) \end{bmatrix}$$

From this, it follows that:

$$(-1)^1 \Delta_1(x, y, z) = 2\lambda_x$$
$$(-1)^2 \Delta_2(x, y, z) = 4\lambda_x \lambda_y - (\lambda_x a_{xy} - \lambda_y a_{yx})^2$$
$$(-1)^3 \Delta_3(x, y, z) = 4\lambda_x \lambda_y \lambda_z (2r_z - 6z) - (\lambda_x a_{xy} - \lambda_y a_{yx})^2 \lambda_z (2r_z - 6z)$$
$$+ 2\lambda_y (\lambda_x a_{xz} - \lambda_z a_{zx})^2$$

This yields the concavity conditions:

$$\lambda_x \geq 0$$
$$4\lambda_x \lambda_y \geq (\lambda_x a_{xy} - \lambda_y a_{yx})^2$$
$$\lambda_z(6z - 2r_z)(4\lambda_x \lambda_y - (\lambda_x a_{xy} - \lambda_y a_{yx})^2) \geq 2\lambda_y(\lambda_x a_{xz} - \lambda_z a_{zx})^2$$

Evaluating Projects and Assessing Sustainable Development in Imperfect Economies

KENNETH J. ARROW[1], PARTHA DASGUPTA[2] and KARL-GÖRAN MÄLER[3]
[1] *Stanford Research Initiative on the Environment, the Economy, and Sustainable Welfare, Stanford University, Stanford;* [2] *Faculty of Economics, University of Cambridge and St. John's College, Cambridge;* [3] *Beijer International Institute of Ecological Economics, Stockholm*

Abstract. We are interested in three related questions: (1) How should accounting prices be estimated? (2) How should we evaluate policy change in an imperfect economy? (3) How can we check whether intergenerational well-being will be sustained along a projected economic programme? We do not presume that the economy is convex, nor do we assume that the government optimizes on behalf of its citizens. We show that the same set of accounting prices should be used both for policy evaluation and for assessing whether or not intergenerational welfare along a given economic path will be sustained. We also show that a comprehensive measure of wealth, computed in terms of the accounting prices, can be used as an index for problems (2) and (3) above. The remainder of the paper is concerned with rules for estimating the accounting prices of several specific environmental natural resources, transacted in a few well known economic institutions.

Key words: accounting prices, bifurcation points, cost-benefit analysis, environmental Kuznets curve, genuine investment, genuine wealth, hysteresis, intertemporal welfare, irreversibility, non-convexity, Pontryagin Principle, population growth, property rights, separatrix, social discount rate, structural stability, sustainable development, technological change, thresholds

1. Introduction

In several recent publications, it has been shown that there is a wealth like measure that can serve as an index of intergenerational welfare. The index enables one (a) to check whether welfare will be sustained along an economic forecast, and (b) to conduct social cost-benefit analysis of policy reforms (e.g., investment projects). Excepting under special circumstances, however, the index in question is not wealth itself, but an adaptation of wealth. Interestingly, the results do not require the economy to be convex, nor do they require the assumption that the government optimizes on behalf of its citizens.[1]

An economy's wealth is the worth of its capital assets. As is widely recognised today, the list of assets should include not only manufactured capital, but also human capital (health, knowledge, and skills), and natural capital. Formally, an economy's wealth is a linear combination of its capital stocks, the weights awarded to the stocks being the latter's accounting prices.

The term accounting prices was used originally in the literature on economic planning (Tinbergen 1954). The underlying presumption there was that governments are intent on maximizing social welfare. Public investment criteria were subsequently developed for economies enjoying good governance (Little and Mirrlees 1968, 1974; Arrow and Kurz 1970). In its turn the now-extensive literature exploring various concepts of sustainable development has also been directed at societies where governments choose policies so as to maximize intergenerational welfare.[2]

Sustainability is different from optimality. To ask whether collective well-being is sustained along an economic forecast is to ask, roughly speaking, whether the economy's production possibility set is growing. The concept of sustainability is useful for judging the performance of economies where the government, whether by design or incompetence, does not choose policies that maximise intergenerational welfare. One can argue, therefore, that the term "sustainable development" acquires particular bite when it is put to work in *imperfect economies*, that is, economies suffering from weak, or even bad, governance. Recently the theory of intertemporal welfare indices has been extended to such economies.[3] The theory's reach therefore now extends to actual economies. The theory has also been put to use in a valuable paper by Hamilton and Clemens (1999) for judging whether in the recent past countries have invested sufficiently to expand their productive bases.[4] Among the resources making up natural capital, only commercial forests, oil and minerals, and the atmosphere as a sink for carbon dioxide were included in the Hamilton-Clemens work. Not included were water resources, forests as agents of carbon sequestration, fisheries, air and water pollutants, soil, and biodiversity. Nor were discoveries of oil and mineral reserves taken into account. Moreover, there is a certain awkwardness in several of the steps Hamilton and Clemens took when estimating changes in the worth of an economy's capital assets. Our aim in this paper is to clarify a number of issues that arise in putting the theory of welfare indices to practical use. It is our hope that the findings documented here will prove useful in future empirical work.

We are interested in three related questions: (1) How should accounting prices be estimated? (2) How should we evaluate policy change in an imperfect economy? (3) How can we check whether intergenerational well-being will be sustained along a projected economic programme?

For simplicity, we confine our analysis until Section 14 to a deterministic world. In Section 2 we rehearse the basic theory.[5] We prove that the same set of accounting prices should be used both for policy evaluation and for assessing whether or not intergenerational welfare along a given economic path will be sustained. We also show that a comprehensive measure of wealth, computed in terms of the accounting prices, can be used as an index for problems (2) and (3) above. These results do not require that the economy be convex, nor do they depend on the assumption that the government optimizes on behalf of its citizens subject to constraints.

In Section 3 we use the Ramsey-Solow model of national saving in a convex economy to illustrate the theory. In Section 4 we show that the theory can be put to use in non-convex economies by studying a particular class of ecosystems, namely, shallow lakes. The remainder of the paper is concerned with rules for estimating the accounting prices of specific environmental natural resources, transacted in a few well known economic institutions.

In order to make our findings easily accessible for empirical work, we report our findings as a catalogue of results. Rules for estimating accounting prices of exhaustible natural resources under both free and restricted entry are derived in Section 5. In Section 6 we show how expenditure toward the discovery of new deposits ought to be incorporated in national accounts. Section 7 develops methods for including forest depletion; and in Section 8 we show how the production of human capital could be taken into account. In Section 9 we study the valuation of global public goods.

If an economy were to face exogenous movements in certain variables, its dynamics would not be autonomous in time. Non-autonomy in time introduces additional problems for the construction of the required welfare index, in that the wealth measure requires to be augmented. Exogenous growth in factor productivities, for example, is a potential reason for non-autonomous dynamics. In Section 10 we show that by suitably redefining variables, it is often possible to transform a non-autonomous economic system into one that is autonomous. But such helpful transformations are not available in many other cases. In Section 11 we show that the required welfare index can nevertheless be constructed, by studying a small country exporting an exhaustible natural resource at a price that is time-dependent. The way defensive expenditure against pollution ought to be included in national accounts is discussed in Section 12.

The theory developed upto and including Section 12 assumes that population remains constant. In Section 13 we extend the theory to cover population change.[6] In Section 14 we show how future uncertainty in commodity transformation possibilities can be incorporated. Section 15 contains concluding remarks.

2. The Basic Model

2.1. PRELIMINARIES

We assume that the economy is closed. Time is continuous and is denoted variously by τ and t (τ, t $\geq$ 0). The horizon is taken to be infinite. For simplicity of exposition, we aggregate consumption into a single consumption good, C, and let **R** denote a vector of resource flows (e.g., rates of extraction of natural resources, expenditure on education and health). Labour is supplied inelastically and is normalised to be unity. Intergenerational welfare (henceforth, "social welfare") at t ($\geq$ 0) is taken to be of the Ramsey-Koopmans form,

$$W_t = \int_t^\infty U(C_\tau)e^{-\delta(\tau-t)}d\tau, \qquad (\delta > 0), \tag{1}$$

where the utility function, U(C), is strictly concave and monotonically increasing.

The state of the economy is represented by the vector **K**, where **K** is a comprehensive list of capital assets. The economy under study faces not only technological and ecological constraints, but also a wide variety of institutional constraints. By the economy's "institutions" we mean market structures, property rights, tax rates, non-market arrangements for credit, insurance, and common property resources, the character of various levels of government, and so forth. We do *not* assume that the government is necessarily bent on maximizing social welfare subject to constraints. It could be that the government is predatory, or is at best neglectful, and has objectives of its own that are not congruent with social welfare. Nor do we imagine institutions to be unchanging over time. What we do assume is that institutions coevolve with the state of the economy (**K**) in ways that are understood. It is no doubt a truism that social and political institutions influence the evolution of the state of an economy, but it has also been argued by political scientists (Lipset 1959) that the state of an economy (**K**) influences the evolution of social and political institutions. The theory we develop below accommodates this mutual influence.

Let $\{C_\tau, \mathbf{R}_\tau, \mathbf{K}_\tau\}_t^\infty$ be an economic programme from t to ∞. Given technological possibilities, resource availabilities, and the dynamics of the ecological-economic system, the decisions made by individual agents and consecutive governments from t onwards will determine C_τ, $\mathbf{R}_\tau$, and $\mathbf{K}_\tau$ – for $\tau \geq t$ – as functions of $\mathbf{K}_t$, τ, and t. Thus let $f(\mathbf{K}_t, \tau, t)$, $\mathbf{g}(\mathbf{K}_t, \tau, t)$, and $\mathbf{h}(\mathbf{K}_t, \tau, t)$, respectively, be consumption, the vector of resource flows, and the vector of capital assets at date τ ($\geq t$) if $\mathbf{K}_t$ is the vector of capital assets at t. Now write

$$(\boldsymbol{\xi}_\tau)_t^\infty \equiv \{C_\tau, \mathbf{R}_\tau, \mathbf{K}_\tau\}_t^\infty, \qquad \text{for } t \geq 0. \tag{2}$$

Let $\{t, \mathbf{K}_t\}$ denote the set of possible t and $\mathbf{K}_t$ pairs, and $\{(\boldsymbol{\xi}_\tau)_t^\infty\}$ the set of economic programmes from t to infinity.

Definition 1

A *resource allocation mechanism*, α, is a (many-one) mapping

$$\alpha : \{t, \mathbf{K}_t\} \to \{(\boldsymbol{\xi}_\tau)_t^\infty\}. \tag{3}$$

It bears emphasis that we do *not* assume that α maps $\{t, \mathbf{K}_t\}$ into to optimum economic programmes (starting at t), nor even that it maps $\{t, \mathbf{K}_t\}$ into efficient programmes (starting at t). The following analysis is valid even if α is riddled with economic distortions and inequities. Nor do we assume, in defining α, that the economy's institutions are fixed. If institutions and the state of the economy were known to coevolve, that coevolution would be reflected in α. Note too that we do *not* assume commodity transformation possibility sets to be convex. This is significant, because ecological processes involve transformation possibility sets that are frequently non-convex; displaying, for example, threshold effects. The reason we are able to accommodate non-convex production structures is that we are developing welfare economics in imperfect economies: we assume that the government (rather, some honest agency in government) seeks only to institute policy reform. For an optimizing government the matter would be different. As the Second Fundamental Theorem of Welfare Economics makes clear, production structures need to be convex if the optimum allocation is to be decentralized.

Definition 2
α is *time-autonomous* (henceforth *autonomous*) if for all $\tau \geq t$, ξ_τ is a function solely of $\mathbf{K}_t$ and $(\tau - t)$.

Notice that if α is autonomous, economic variables at date τ ($\geq t$) are functions of $\mathbf{K}_t$ and $(\tau - t)$ only. α would be non-autonomous if, for example, knowledge or the terms of trade (for a trading economy) were to change exogenously over time. In certain cases exogenous changes in population size would mean that α is not autonomous. However, by suitably redefining state variables, non-autonomous resource allocation mechanisms can sometimes be mapped into autonomous mechanisms (Sections 10 and 13).

Definition 3
α is *time-consistent* if

$$\mathbf{h}(\mathbf{K}_{\tau'}, \tau'', \tau') = \mathbf{h}(\mathbf{K}_t, \tau'', t), \qquad \text{for all } \tau'', \tau', \text{ and } t. \tag{4}$$

Time-consistency implies a weak form of rationality. An autonomous resource allocation mechanism, however, has little to do with rationality; it has to do with the influence of external factors (e.g., whether trade prices are changing autonomously). In what follows, we assume that α is time-consistent.

Definition 4
The *value function* reflects social welfare (equation (1)) as a function of initial capital stocks and the resource allocation mechanism. We write this as

$$W_t = V(\mathbf{K}_t, \alpha, t). \tag{5}$$

In what follows, we will often write $V(\mathbf{K}_t, \alpha, t) = V_t$.
Let K_i be the ith capital stock. We assume that V is differentiable in $\mathbf{K}$.[7]

Definition 5
The *accounting price*, p_{it}, of the ith capital stock is defined as

$$p_{it} = \partial V(\mathbf{K}_t, \alpha, t)/\partial K_{it} \equiv \partial V_t/\partial K_{it}. \tag{6}$$

Note that accounting prices are defined in terms of hypothetical perturbations to an economic forecast. Specifically, the accounting price of a capital asset is the present discounted value of the perturbations to U that would arise from a marginal increase in the quantity of the asset. Given the resource allocation mechanism, accounting prices at t are functions of $\mathbf{K}_t$, and possibly of t as well (i.e., $p_{it} = p_i(\mathbf{K}_t, t)$). The prices depend also on the extent to which various capital assets are substitutable for one another. It should be noted that accounting prices of private "goods" can be negative if property rights are dysfunctional, such as those that lead to the tragedy of the commons. Note too that if α is autonomous, accounting prices are not explicit functions of time, and so, $p_{it} = p_i(\mathbf{K}_t)$.

2.2. MARGINAL RATES OF SUBSTITUTION VS MARKET OBSERVABLES

Using (1) and (6), it can be shown that, if α is autonomous, p_{it} satisfies the dynamical equation,

$$dp_{it}/dt = \delta p_{it} - U'(C_t)\partial C_t/\partial K_{it} - \sum_j p_{jt}\partial(dK_{jt}/dt)/\partial K_{it}. \tag{7}$$

(7) reduces to Pontryagin equations for co-state variables in the case where α is an optimum resource allocation mechanism. In any event, we show below that, in order to study the evolution of accounting prices under simple resource allocation mechanisms, it is often easier to work directly with (6).

From (6) it also follows that accounting price ratios (p_{it}/p_{jt}, $p_{i\tau}/p_{it}$, and consumption discount rates (see below)) are defined as marginal social rates of substitution between goods. In an economy where the government maximizes social welfare, marginal rates of substitution among goods and services equal their corresponding marginal rates of transformation. As the latter are observable in market economies (e.g., border prices for traded goods in an open economy), accounting prices are frequently defined in terms of marginal rates of transformation among goods and services. However, marginal rates of substitution in imperfect economies do not necessarily equal the corresponding marginal rates of transformation. A distinction therefore needs to be made between the ingredients of social welfare and "market observables". Using market observables to infer social welfare can be misleading in imperfect economies. That we may have to be explicit about welfare parameters (e.g., δ and the elasticity of $U'(C)$) in order to estimate marginal rates of substitution in imperfect economies is not an argument for pretending that the economies in question are not imperfect after all. In principle it could be hugely misleading to use the theory of optimum control to justify an exclusive interest in market observables.

2.3. GENUINE INVESTMENT AS A MEASURE OF SUSTAINABLE DEVELOPMENT

IUCN (1980) and World Commission (1987) introduced the concept of sustainable development. The latter publication defined sustainable development to be "... development that meets the needs of the present without compromising the ability of future generations to meet their own needs" (World Commission 1987, p. 43).

Several formulations are consistent with this phrase. But the underlying idea is straightforward enough: we seek a measure that would enable us to judge whether an economy's production possibility set is, in a loose sense, growing. Our analysis is based on an interpretation of sustainability that is based on the maintainence of social welfare, rather than on the maintainenance of the economy's productive base. We then show that the requirement that economic development be sustainable implies, and is implied by, the requirement that the economy's productive base be maintained (Theorems 1–3). These results give intellectual support for the definition of sustainability we adopt here.[8]

Definition 6
The economic programme $\{C_t, \mathbf{R}_t, \mathbf{K}_t\}_0^\infty$ corresponds to a *sustainable development path* at t if $dV_t/dt \geq 0$.[9]

Notice that the above criterion does not attempt to identify a unique economic programme. In principle any number of technologically and ecologically feasible economic programmes could satisfy the criterion. On the other hand, if substitution possibilities among capital assets are severely limited and technological advances are unlikely to occur, it could be that there is no sustainable economic programme open to an economy. Furthermore, even if the government were bent on optimising social welfare, the chosen programme would not correspond to a sustainable path if the utility discount rate, δ, were too high. It could also be that along an optimum path social welfare declines for a period and then increases thereafter, in which case the optimum programme does not correspond to a sustainable path locally, but does so in the long run.[10]

Optimality and sustainability are thus different notions. The concept of sustainability helps us to better understand the character of economic programmes, and is particularly useful for judging the performance of imperfect economies.

We may now state

Theorem 1

$$dV_t/dt = \sum_i p_{it} dK_{it}/dt + \partial V_t/\partial t. \qquad (8)$$

The proof follows directly from equations (5) and (6).

Definition 6
The accounting value of the rate of change in the stocks of capital assets is called *genuine investment*.

If α is autonomous, then $\partial V_t/\partial t = 0$, and so, from equation (8) we have,

Theorem 2

If α is autonomous, then

$$dV_t/dt = \sum_i p_{it}dK_{it}/dt.\text{[11]} \tag{9}$$

Equation (9) states that at each date the rate of change in social welfare equals genuine investment. Theorem 2 gives a local measure of sustainability. Integrating (9) yields a non-local measure:

Theorem 3

If α is autonomous, for all $T \geq 0$,

$$V_t - V_0 = \sum_i \left[p_{it}K_{it} - p_{i0}K_{i0}\right] - \int_0^T \left[\sum_i (dp_{i\tau}/d\tau)K_{i\tau}\right] d\tau. \tag{10}$$

Equation (10) shows that in assessing whether or not social welfare has increased between two dates, the "capital gains" on the assets that have accrued over the interval should be deducted from the difference in wealth between the dates.

Each of Theorems 1, 2 and 3 is an equivalence result. None says whether α gives rise to an economic programme along which social welfare is sustained. For example, it can be that an economy is incapable of achieving a sustainable development path, owing to scarcity of resources, limited substitution possibilities among capital assets, or whatever. Or it can be that although the economy is in principle capable of achieving a sustainable development path, social welfare is unsustainable along the path that has been forecast because of bad government policies. Or it can be that α is optimal, but that because the chosen utility discount rate is large, social welfare is not sustained along the optimum economic programme. Or it can be that along an optimum path social welfare declines for a period and then increases thereafter.

2.4. WHAT ELSE DOES GENUINE INVESTMENT MEASURE?

Genuine investment is related to changes in future consumption brought about by it. Imagine that the capital base at t is not $\mathbf{K}_t$ but $\mathbf{K}_t + \Delta\mathbf{K}_t$, where as before, Δ is an operator signifying a small difference. In the obvious notation,

$$V(\alpha, \mathbf{K}_t + \Delta\mathbf{K}_t) - V(\alpha, \mathbf{K}_t) \approx \int_t^\infty U'(C_\tau)\Delta(C_\tau)e^{-\delta(\tau-t)}d\tau. \tag{11}$$

Now suppose that at t there is a small change in α, but only for a brief moment, Δt, after which the resource allocation mechanism reverts back to α. We write the increment in the capital base at $t + \Delta t$ consequent upon the brief increase in genuine investment as $\Delta\mathbf{K}_t$. So $\Delta\mathbf{K}_t$ is the consequence of an increase in genuine investment at t and $(\mathbf{K}_{t+\Delta t} + \Delta\mathbf{K}_t)$ is the resulting capital base at $t + \Delta t$. Let Δt tend to zero. From equation (11) we obtain

Theorem 4
Genuine investment measures the present discounted value of the changes to consumption services brought about by it.[12]

2.5. PROJECT EVALUATION CRITERIA

Theorem 4 provides a criterion for social cost-benefit analysis of policy reforms. Imagine that even though the government does not optimize, it can bring about small changes to the economy by altering the existing resource allocation mechanism in minor ways. The perturbation in question could be small adjustments to the prevailing structure of taxes for a short while, or it could be minor alterations to the existing set of property rights for a brief period, or it could be a small public investment project. Call any such perturbation a "policy reform".

Consider as an example an investment project. It can be viewed as a perturbation to the resource allocation mechanism α for a brief period (the lifetime of the project), after which the mechanism reverts back to its earlier form. We consider projects that are small relative to the size of the economy. How should they be evaluated?

For simplicity of exposition, we suppose there is a single manufactured capital good (K) and a single extractive natural resource (S). The rate of extraction is denoted by R. Let the project's lifetime be the period [0, T]. Denote the project's output and inputs at t by the vector $(\Delta Y_t, \Delta L_t, \Delta K_t, \Delta R_t)$. We imagine that if the project is accepted, the project manager would rent ΔK_t at t for the period t to t + Δt.[13]

The project's acceptance would perturb consumption under α. Let the perturbation at t (≥ 0) be $\tilde{\Delta} C_t$. It would affect U_t by the amount $U'(C_t)\tilde{\Delta} C_t$. However, because the perturbation includes all "general equilibrium effects", it would be tiresome if the project evaluator were required to estimate $\tilde{\Delta} C_t$ for every project that came up for consideration. Accounting prices are useful because they enable project evaluators to estimate $\tilde{\Delta} C_t$ indirectly, which means that they do not have to go beyond project data in order to evaluate projects. Now, it is most unlikely that consumption and investment have the same accounting price in an imperfect economy. So we divide ΔY_t into two parts: changes in consumption and in investment in manufactured capital. Denote them as ΔC_t and $\Delta(dK/dt)$, respectively.

U is the unit of account.[14] Let w_t denote the accounting wage rate. Next, let q_t be the accounting price of the extractive resource input of the project and λ_t the social cost of borrowing capital (i.e., $\lambda_t = \delta - [dp_t/dt]/p_t$).[15]

From the definition of accounting prices, it follows that:

$$\int_0^\infty U'(C_\tau)\tilde{\Delta} C_\tau e^{-\delta\tau} d\tau = \qquad (12)$$
$$\int_0^T (U'(C_\tau)\Delta C_\tau + p_\tau \Delta(dK_\tau/d\tau) - w_\tau \Delta L_\tau - \lambda_\tau p_\tau \Delta K_\tau - q_\tau \Delta R_\tau)e^{-\delta\tau} d\tau.$$

But the RHS of (12) is the present discounted value of social profits from the project (in utility numeraire). Moreover, $\int_0^\infty U'(C_\tau)\tilde{\Delta}C_\tau e^{-\delta\tau}d\tau = \Delta V_0$, the latter being the change in social welfare if the project were accepted. We may therefore write (12) as,

$$\Delta V_0 = \int_0^T (U'(C_\tau)\Delta C_\tau + p_\tau \Delta(dK_\tau/d\tau) - w_\tau \Delta L_\tau - \lambda_\tau p_\tau \Delta K_\tau - q_\tau \Delta R_\tau)e^{-\delta\tau}d\tau. \quad (13)$$

Equation (13) leads to the well-known criterion for project evaluation:

Theorem 5
A project should be accepted if and only if the present discounted value of its social profits is positive.

2.6. NUMERAIRE

So far we have taken utility to be the unit of account. In applied welfare economics, however, it has been found useful to express benefits and costs in terms of current consumption. It will pay to review the way the theory being developed here can be recast in consumption numeraire. For simplicity of exposition, assume that there is a single commodity, that is, an all-purpose durable good that can be consumed or reinvested for its own accumulation. Assume too that the elasticity of marginal utility is a constant, η. Define $\bar{p}_t$ to be the accounting price of the asset at t in terms of consumption at t; that is,

$$\bar{p}_t = p_t/U'(C_t). \quad (14)$$

It follows from (14) that,

$$(d\bar{p}_t/dt)/\bar{p}_t = (dp_t/dt)/p_t + \eta(dC_t/dt)/C_t. \quad (15)$$

Let ρ_t be the social rate of discount in consumption numeraire. ρ_t is sometimes referred to as the consumption rate of interest (Little and Mirrlees 1974). From (1),

$$\rho_t = \delta + \eta(dC_t/dt)/C_t.^{16} \quad (16)$$

Using (16) in (15) we obtain the relationship between the asset's prices in the two units of account:

$$(d\bar{p}_t/dt)/\bar{p}_t = (dp_t/dt)/p_t + \rho_t - \delta.^{17} \quad (17)$$

2.7. INTRAGENERATIONAL DISTRIBUTION

The distribution of well-being within a generation has been ignored so far. Theoretically it is not difficult to include this. If there are N people in each generation and person j consumes C_j, her welfare would be $U(C_j)$.[18] A simple way to express *intra*generational welfare would be to "concavify" U. Let G be a strictly concave, increasing function of real numbers. We may then express intragenerational welfare as $\sum_j(G(U(C_j)))$. Some people would be well-off, others badly-off. The formulation ensures that at the margin, the well-being of someone who is badly off is awarded greater weight than that of someone well-off.

The social worth of consumption services (C) depends on who gets what. To accommodate this idea, we have to enlarge the set of commodities so as to distinguish, at the margin, a good consumed or supplied by one person from that same good consumed or supplied by another. Thus, a piece of clothing worn by a poor person should be regarded as a different commodity from that same type of clothing worn by someone who is rich. With this re-interpretation of goods and services, the results we have obtained continue to hold.

Relatedly, we should note that the connection between rural poverty in the world's poorest regions and the state of the local ecosystems is a close one. When wetlands, inland and coastal fisheries, woodlands, forests, ponds and lakes, and grazing fields are damaged (owing, say, to agricultural encroachment, or urban extensions, or the construction of large dams, or organizational failure at the village level), traditional dwellers suffer. For them – and they are among the poorest in society – there are frequently no alternative source of livelihood. In contrast, for rich eco-tourists or importers of primary products, there is something else, often somewhere else, which means that there are alternatives. Whether or not there are substitutes for a particular resource is therefore not only a technological matter, nor a mere matter of consumer taste: among poor people location can matter too. The poorest of the poor experience non-convexities in a way the rich do not. Even the range between a need and a luxury is context-ridden. Macroeconomic reasoning glosses over the heterogeneity of Earth's resources and the diverse uses to which they are put – by people residing at the site and by those elsewhere.[19]

3. Illustration, 1: A Convex Production Economy

It will prove useful to illustrate the theory by means of a simple example, based on Ramsey (1928) and Solow (1956). As in Section 2.6, imagine that there is an all-purpose durable good, whose stock at t is K_t (≥ 0). The good can be consumed or reinvested for its own accumulation. There are no other assets. Write output (GNP) as Y. Technology is linear. So $Y = \mu K$, where $\mu > 0$. μ is the output-wealth ratio. GNP at t is $Y_t = \mu K_t$.

Imagine that a constant proportion of GNP is saved at each moment. There is no presumption though that the saving rate is optimum; rather, it is a behavioural characteristic of consumers, reflecting their response to an imperfect credit market. Other than this imperfection, the economy is assumed to function well. At each moment expectations are fulfilled and all markets other than the credit market clear. This defines the resource allocation mechanism, α. Clearly, α is autonomous in time. We now characterise α explicitly.

Let the saving ratio be s ($0 < s < 1$). Write aggregate consumption as C_t. Therefore,

$$C_t = (1-s)Y_t = (1-s)\mu K_t. \tag{18}$$

Capital is assumed to depreciate at a constant rate γ (> 0). Genuine investment is therefore,

$$dK_t/dt = (s\mu - \gamma)K_t. \tag{19}$$

K_0 is the initial capital stock. The economy grows if $s\mu > \gamma$, and shrinks if $s\mu < \gamma$. To obtain a feel for orders of magnitude, suppose $\gamma = 0.05$ and $\mu = 0.25$. The economy grows if $s > 0.2$, and shrinks if $s < 0.2$.

Integrating (19), we obtain,

$$K_\tau = K_t e^{(s\mu-\gamma)(\tau-t)}, \qquad \tau \geq t \geq 0, \tag{20}$$

from which it follows that,

$$C_\tau = (1-s)\mu K_\tau = (1-s)\mu K_t e^{(s\mu-\gamma)(\tau-t)}, \qquad \tau \geq t \geq 0. \tag{21}$$

If the capital stock was chosen as numeraire, wealth would be K_t, and NNP would be $(\mu - \gamma)K_t$. Each of wealth, GNP, NNP, consumption and genuine investment expands at the exponential rate $(s\mu - \gamma)$ if $s\mu > \gamma$; they all contract at the exponential rate $(\gamma - s\mu)$ if $s\mu < \gamma$. We have introduced capital depreciation into the example so as to provide a whiff (albeit an artificial whiff) of a key idea, that even if consumption is less than GNP, wealth declines when genuine investment is negative. Wealth declines when consumption exceeds NNP.

Current utility is $U(C_t)$. Consider the form

$$U(C) = -C^{-(\eta-1)}, \qquad \text{where } \eta > 1.^{20} \tag{22}$$

η is the elasticity of marginal utility and δ is the social rate of discount if utility is numeraire. Let ρ_t be the social rate of discount if consumption is the unit of account. It follows that

$$\rho_t = \delta + \eta(dC_t/dt)/C_t = \delta + \eta(s\mu - \gamma). \tag{23}$$

The sign of ρ_t depends upon the resource allocation mechanism α. In particular, ρ_t can be negative. To see why, suppose the unit of time is a year, $\delta = 0.03$, $\gamma = 0.04$, $s = 0.10$, $\eta = 2$, and $\mu = 0.20$. Then $\eta(dC_t/dt)/C_t = -0.04$ per year, and (23) says that $\rho_t = -0.01$ per year.[21]

Social welfare at t is,

$$V_t = \int_t^\infty U(C_\tau)e^{-\delta(\tau-t)}d\tau. \tag{24}$$

Using (21) and (22) in (24), we have:

$$V_t = -[(1-s)\mu K_t]^{-(\eta-1)} \int_t^\infty e^{-[(\eta-1)(s\mu-\gamma)+\delta](\tau-t)}d\tau,$$

or, assuming that $[(\eta - 1)(s\mu - \gamma) + \delta] > 0$,

$$V_t = -[(1 - s)\mu K_t]^{-(\eta-1)}/[(\eta - 1)(s\mu - \gamma) + \delta]. \tag{25}$$

V is differentiable in K everywhere. Moreover, $\partial V_t/\partial t = 0$. Equations (20) and (25) confirm Theorem 1.[22]

We turn now to accounting prices.

(i) UTILITY NUMERAIRE

Begin by taking utility to be numeraire. Let p_t be the accounting price of capital. Now

$$p_t \equiv \partial V_t/\partial K_t = \int_t^\infty U'(C_\tau)[\partial C_\tau/\partial K_t]e^{-\delta(\tau-t)}d\tau. \tag{26}$$

Using (25) in (26) we have,

$$p_t = (\eta - 1)[(1 - s)\mu]^{-(\eta-1)}K_t^{-\eta}/[(\eta - 1)(s\mu - \gamma) + \delta]. \tag{27}$$

Using equations (20), (21), (25), and (27) it is simple to check that $p_t \neq U'(C_t)$, except when $s = (\mu + (\eta - 1)\gamma - \delta)/\mu\eta$. Let s* be the optimum saving rate. From equation (25) we have,

$$s^* = (\mu + (\eta - 1)\gamma - \delta)/\mu\eta. \tag{28}$$

Note that $p_t < U'(C_t)$ if $s > s^*$, which means there is excessive saving. Conversely, $p_t > U'(C_t)$ if $s < s^*$, which means there is excessive consumption.

(ii) CONSUMPTION NUMERAIRE

Write $\bar{p}_t = p_t/U'(C_t)$. \hfill (29)

Using (26) in (29) yields

$$\bar{p}_t = \int_t^\infty [U'(C_\tau)/U'(C_t)][\partial C_\tau/\partial K_t]e^{-\delta(\tau-t)}d\tau. \tag{30}$$

Now use (21), (22) and (30) to obtain

$$\bar{p}_t = \int_t^\infty (1 - s)\mu e^{(-\rho+(s\mu-\gamma))(\tau-t)}d\tau, \tag{31}$$

where $\rho = \delta + \eta(s\mu - \gamma)$.

From (31) we have

$$\bar{p}_t = (1 - s)\mu/(\rho - (s\mu - \gamma)). \tag{32}$$

Observe that $\bar{p}_t > 1$ (resp. < 1) if $s < s^*$ (resp. $> s^*$).[23]

In order to obtain a sense of orders of magnitude, suppose $\eta = 2$, $\mu = 0.20$, $\gamma = 0.05$, and $\delta = 0$. From (28) we have s* = 0.625. Now imagine that s = 0.40 (by Ramsey's criterion, this is undersaving!). Using (23) we have $\rho = 0.06$ per unit of time. So (32) reduces to $\bar{p}_t = 4$. In other words, a saving rate that is approximately 30 percent short of the optimum corresponds to a high figure for the accounting price of investment: investment should be valued four times consumption.

Although intergenerational equity is nearly always discussed in terms of the rate at which future well-being is discounted (see e.g., Portney and Bryant 1998), equity would be more appropriately discussed in terms of the curvature of U. Let the unit of time be a year. Suppose $\gamma = 0$, $\delta = 0.02$, and $\mu = 0.32$. Consider two alternative values of η: 25 and 50. It is simple to confirm that s* = 0.038 if $\eta = 25$ and s* = 0.019 if $\eta = 50$. Intergenerational equity in both consumption and welfare (the latter is a concave function of the former) can be increased indefinitely by making η larger and larger: C_t becomes "flatter" as η is increased. In the limit, as η goes to infinity, s* tends to γ (equation (28)), which reflects the Rawlsian maxi-min consumption as applied to the intergenerational context.[24]

4. Illustration, 2: A Non-convex Ecosystem

The Ramsey-Solow economy discussed above is convex. In this section we confirm that the theory presented in Section 2 can be applied to non-convex economies. We do this by studying a model of shallow lakes.[25]

A key determinant of the overall state of a shallow lake is phosphorus, which is a necessary nutrient for such ecological services in the lake as those that provide a habitat for fish populations. But at high levels of concentration phosphorus is a pollutant, causing as it does increased plant growth, algae blooms, decrease in water transparency, bad odour, oxygen depletion, and fish kills. Thus, the state of a lake can be taken to be the quantity of phosphorus in the water column, which we denote by a scalar, S.

The rate of phosphorus inflow into a lake is a byproduct of agriculture in the watershed (e.g., as fertilizer runoff from farms). We bring these considerations together and postulate that current utility is a strictly concave and twice differential function U(C, S), where U is an increasing function of phosphorus inflow, C. Imagine next that phosphorus has a deleterious effect on the lake at all levels of concentration (and not just at high levels of concentration); which is to say that U is a decreasing function of S for all S. This assumption brings into sharp relief those economic problems where a produced good has positive social worth as a flow, even though it is a pollutant as a stock.

Social welfare at t is

$$V(S_t) = \int_t^\infty U(C_\tau, S_\tau) e^{-\delta(\tau-t)} d\tau, \text{ where } U_S < 0 \text{ and } U_C > 0.$$

4.1. CONSTANT PHOSPHORUS INFLOW

For simplicity of exposition, we suppose in what follows that

$$U(C, S) = \log C - hS^2, \quad h > 0. \tag{33}$$

Consider the case where the resource allocation mechanism for phosphorus inflow is such that C_t is a constant, say $\bar{C}$. Studies have confirmed that there is a feedback of phosphorus from bottom sediments when the density of algae in the lake is large. This feedback is reflected in the form of recycling – from sediment to the water column. Experiments suggest that the recycling rate, R, is a sigmoid function of S. A simple form of the relationship is,

$$R_t = bS_t^2/(1 + S_t^2), \quad \text{where } b > 0. \tag{34}$$

The rate of input of phosphorus into the water column is therefore $[\bar{C} + bS_t^2/(1 + S_t^2)]$.

However, phosphorus is depleted from the water column owing to sedimentation and water outflow. Assuming that the rate of loss is proportional to S, say γS ($\gamma > 0$), the phosphorus content in the lake's water column is governed by the equation,

$$dS_t/dt = \bar{C} + bS_t^2/(1 + S_t^2) - \gamma S_t. \tag{35}$$

For a range of parameter values $\bar{C}$, b, and γ, the curves $[\bar{C} + bS^2/(1 + S^2)]$ and γS intersect at three points. This is shown in Figure 1. The upper and lower intersects, S_3 and S_1, are stable stationary points of (35), whereas the intermediate intersect, S_2, is unstable. Thus, S_2 is the unique separatrix of the dynamical system. S_3 and S_1 should be thought of as eutrophic and oligotrophic states, respectively. Thus, given S_t, the resource allocation mechanism, α, governing the lake's quality can be expressed as,

$$dS_\tau/d\tau = \bar{C} + bS_\tau^2/(1 + S_\tau^2) - \gamma S_\tau, \quad \tau \geq t. \tag{36}$$

Clearly, α is autonomous and time consistent. It is simple to confirm that V(S) is differentiable in S everywhere, excepting S_2. It is simple to confirm as well that, although V(S) is discontinuous at S_2, it possesses both right- and left-hand derivatives there. We can therefore define the accounting price of the lake's quality to be $p(S) = \partial V/\partial S$ at all $S \neq S_2$ and apply the theory locally for the purposes of project evaluation and sustainability assessment. It should be noted that because phosphorus is a pollutant in the lake, $p(S) < 0$.[26]

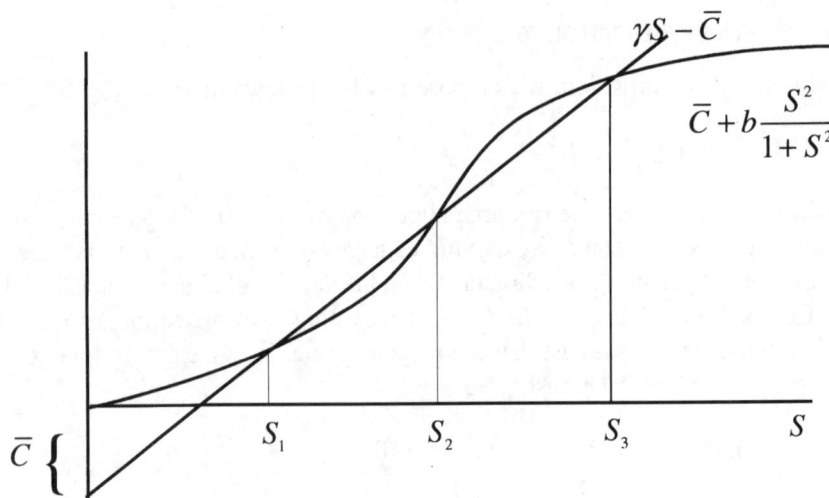

Figure 1. Dynamics of a shallow lake.

4.2. OPTIMUM PHOSPHORUS INFLOW

The resource allocation mechanism defined by (36) reflects an imperfect economy. Brock and Starrett (2003) have studied the optimum resource allocation mechanism. To review their work, we generalize (36). If C_t is the inflow of phosphorus, the lake's dynamics are given by the equation,

$$dS_t/dt = C_t + bS_t^2/(1 + S_t^2) - \gamma S_t, \qquad \text{for } t \geq 0, \tag{37}$$

where S_0 is given as an initial condition.

The problem is to choose $\{C_t\}_0^\infty$ so as to maximize (33), subject to (37).

Clearly, the optimum resource allocation mechanism is both autonomous and time consistent. In what follows, we restrict ourselves to the case where the optimum is an interior one (i.e., $C_t > 0$). Let p_t be the accounting price of phosphorus in the lake. Brock and Starrett confirmed that, for $\{C_t\}_0^\infty$ to be an optimum, it is necessary that C_t and S_t satisfy not only (37), but also the Pontryagin conditions,

$$p_t = -U_C(< 0), \qquad \text{for all } t, \tag{38}$$

and $\quad (dp_t/dt)/p_t = \delta + \gamma - U_S/U_C - 2bS_t/(1 + S_t^2)^2, \qquad \text{for all } t. \tag{39}$

The point therefore is to select p_0 (equivalently, C_0) optimally and allow the dynamical system to evolve in accordance with equations (37)–(39). The authors showed that, in the (p, S) space, equations (37)–(39) can have at most a countable number of stationary points. They studied in detail the class of parameter values for which the number of stationary points is three. They found that two of them (call them S_1 and S_3, with $S_1 < S_3$, corresponding to what could be interpreted to be the oligotrophic and eutrophic state, respectively) are saddle points, while the intermediate point (call it S_2) is a spiral source (i.e., it is unstable).[27] The authors showed that there exists a value of phosphorus stock, $\bar{S}$, such that if $S_0 > \bar{S}$, the optimum programme asymptotes to S_3; but if $S_0 < \bar{S}$, it asymptotes to S_1. In short, history matters.[28] It is easy to confirm that if, by fluke, $S_0 = \bar{S}$, there are two equally desirable optimal programmes, one that asymptotes to S_1, another that asymptotes to S_3. This last property can be shown to imply that V(S), although not differentiable at $\bar{S}$, is continuous at $\bar{S}$ and possesses both left- and right-derivatives. $\bar{S}$ is an endogenously determined separatrix.[29]

Since the optimum resource allocation mechanism is autonomous, we may write by p(S) the optimum policy function. Phosphorus being a pollutant in the lake, we have p(S) < 0. It can be shown that V(S) is differentiable everywhere excepting at $\bar{S}$. It can also be demonstrated that p(S) is discontinuous at $\bar{S}$, but is left- and right-differentiable there. Moreover,

$$p(S) = \partial V/\partial S (< 0), \quad \text{for all } S \neq \bar{S}. \tag{40}$$

Writing by $[p(S)]_{\bar{S}-0}$ (resp., $[p(S)]_{\bar{S}+0}$) the limit of p(S) as S tends to $\bar{S}$ from the left (resp., right), and similarly for $[\partial V/\partial S]_{\bar{S}-0}$ and $[\partial V/\partial S]_{\bar{S}+0}$, it can be shown too that $[p(S)]_{\bar{S}-0} = [\partial V/\partial S]_{\bar{S}-0}$ and $[p(S)]_{\bar{S}+0} = [\partial V/\partial S]_{\bar{S}+0}$. The theory we have outlined in Section 2 is thus applicable to the optimum resource allocation mechanism of this particular non-convex economy.

Having illustrated the theory by means of a three examples, we now proceed to obtain rules for estimating accounting prices. We do this by focussing on specific categories of capital assets and several well known institutional imperfections.

5. Exhaustible Resources: The Closed Economy

Accounting prices of exhaustible resources when depletion rates are optimal have been much studied (e.g., Dasgupta and Heal 1979; see below). What is the structure of their accounting prices when resources are instead common pools?

Two property-rights regimes suggest themselves: open access and restricted entry. They in turn need to be compared to an optimum regime. It is simplest if we avoid a complete capital model. So we resort to a partial equilibrium world: income effects are assumed to be negligible. Let R_t be the quantity extracted at t. Income is the numeraire. Let U(R) be the area under the demand curve below R. So U'(R) is taken to be the market demand function. U is assumed to be an increasing and strictly concave function of R for positive values of R. In order to have a notation that is consistent with the one in the foregoing example, we take the social rate of interest to be an exogenously given constant, ρ. Let S_t be the stock. Then,

$$dS_t/dt = -R_t. \tag{41}$$

5.1. THE OPTIMUM REGIME

In order to construct a benchmark against which imperfect economies can be evaluated, we first study an optimizing economy. Assume that extraction is costless (constant unit extraction cost can be introduced easily). Social welfare at t is,

$$V_t = \int_t^\infty U(R_\tau) e^{-\rho(\tau-t)} d\tau. \tag{42}$$

Let p_t^* denote the accounting price of the resource underground (equivalently, the Hotelling rent, or the optimum depletion charge per unit extracted). We know that

$$dp_t^*/dt = \rho p_t^*. \tag{43}$$

This is the Hotelling Rule. Moreover, optimum extraction, R_t^*, must satisfy the condition,

$$U'(R_t) = p_t^*. \tag{44}$$

Assume that

$$U(R) = -R^{-(\eta-1)}, \quad \text{where } \eta > 1. \tag{45}$$

Then

$$R_t^* = (\rho/\eta) S_0 e^{-\rho t/\eta}. \tag{46}$$

We next consider the two imperfect regimes.

5.2. RESTRICTED ENTRY

For vividness, assume that there are N identical farmers (i, j = 1, 2, ..., N), drawing from an unrechargeable aquifer. Extraction is costless. We model the situation in the following way:[30]

At t, farmer i owns a pool of size S_{it}. Each pool is separated from every other pool by a porous barrier. Water percolates from the pool which is larger to the one which is smaller. Let λ_{ij} (> 0), be the rate at which water diffuses from pool i to pool j. We assume that $\lambda_{ij} = \lambda_{ji}$. Denote by R_{it} the rate at which i draws from his pool. There are then N depletion equations:

$$dS_{it}/dt = \sum_{N-i} \left[\lambda_{ji}(S_{jt} - S_{it})\right] - R_{it}, \tag{47}$$

where "$\sum_{N-i}$" denotes summation over all j other than i.

The payoff function for farmer i at time t is

$$\int_t^\infty U(R_{i\tau}) e^{-\rho(\tau-t)} d\tau. \tag{48}$$

Farmers play non-cooperatively. For tractablity, we study an open loop solution: Farmers are assumed to be naive (when computing his own optimum extraction rates, each takes the others' extraction rates as given).

Let p_{it} be the (spot) personal accounting price of a unit of i's own resource pool. The present value Hamiltonian for i's optimization problem would then be,

$$H_0 = U(R_{it})e^{-\rho t} + \left[\sum_{N-i} \lambda_{ji}(S_{jt} - S_{it}) - R_{it}\right] p_{it}e^{-\rho t}. \tag{49}$$

It follows from (49) that p_{it} obeys the equation,

$$dp_{it}/dt = (\rho + \sum_{N-i} \lambda_{ji})p_{it}. \tag{50}$$

For notational simplicity, assume that $\lambda_{ij} = \lambda$ for all i, j. Then (50) reduces to

$$dp_{it}/dt = (\rho + (N-1)\lambda)p_{it}. \tag{51}$$

Write $(\rho + (N-1)\lambda) = \beta$. We conclude that the rush to extract because of insecure property rights amounts to each extractor using an implicit discount rate, β, which is in excess of the social discount rate ρ.[31] Assume now that the elasticity of demand is a constant, η (> 1). Using (46) and (51), we conclude that the extraction rate from the common pool is

$$R_\tau = (\beta/\eta)S_t e^{-\beta(\tau-t)/\eta}, \qquad \text{for all } \tau \geq t. \tag{52}$$

In order to have a meaningful problem, we take it that $\beta/\eta > \beta - \rho$ (see below).

Let p_t be the resource's (social) accounting price. We know that $p_t = \partial V_t/\partial S_t$. Using (46), it follows that,

$$p_t = \int_t^\infty U'(R_\tau)[\partial R_\tau/\partial S_t]e^{-\rho(\tau-t)}d\tau. \tag{53}$$

Write $\bar{p}_t = p_t/U'(R_t)$. Then (51) and (53) imply

$$\bar{p}_t = \beta/(\beta - \eta(\beta - \rho)) > 1. \tag{54}$$

(Notice that $\bar{p}_t = 1$ if $\beta = \rho$.)

As a numerical illustration, consider the case where $\rho = 0.06$, $\beta = 0.10$, and $\eta = 2$. In this case, $\bar{p}_t = 5$, which reflects a considerable imperfection in the resource allocation mechanism in question: the resource's accounting price is five times its market price.

5.3. OPEN ACCESS

We next study an open-access pool. To have a meaningful problem, we now assume that extraction is costly. For simplicity, let the unit extraction cost be a constant k (> 0). Under open access, Hotelling rents are dissipated completely. Therefore, the equilibrium extraction rate, R_t, is the solution of the equation,

$$U'(R_t) = k. \tag{55}$$

Equation (55) confirms that, for any given level of reserves, there is excessive extraction. Let $\bar{R}$ be the solution of (55). We then have,

$$dS_t/dt = -\bar{R}.$$

Reserves remain positive for a period $T = S/\bar{R}$. Let us normalize utility by setting $U(0) = 0$. It follows that,

$$V_t = \int_t^{(t+S_t/\bar{R})} (U(\bar{R}) - k\bar{R})e^{-\rho(\tau-t)}d\tau. \tag{56}$$

Let p_t be the accounting price of the unextracted resource. Then,

$$p_t = \partial V_t/\partial S_t = [(U(\bar{R}) - k\bar{R})/\bar{R}]e^{-\rho S_t/\bar{R}} > 0. \tag{57}$$

Write $\bar{p}_t = p_t/U'(\bar{R})$, which is the ratio of the resource's shadow price to its unit extraction cost. Then, from (55) and (57),

$$\bar{p}_t = [(U(\bar{R}) - k\bar{R})/k\bar{R}]e^{-\rho S_t/\bar{R}} > 0. \tag{58}$$

(58) resembles a formula proposed by El Serafy (1989) for estimating depletion charges.[32] The charge is positive because an extra unit of water in the aquifer would extend the period of extraction. Notice that $\bar{p}_t$ is bounded above by the ratio of the Marshallian consumer surplus to total extraction cost; furthermore, it increases as the aquifer is depleted and attains its upper bound at the date at which the pool is exhausted. If reserves are large, $\bar{p}_t$ is small, and free access involves no great loss – a familiar result.

What are plausible orders of magnitude? Consider the linear demand function. Assume therefore that

$$U(R) = aR - bR^2, \quad \text{where } a > k \text{ and } b > 0. \tag{59}$$

From (55) and (59),

$$\bar{R} = (a - k)/2b. \tag{60}$$

Substituting (59) and (60) in (58),

$$\bar{p}_t = ((a - k)/2k)e^{-2b\rho S_t/(a-k)}. \tag{61}$$

Equation (61) says that

$$\bar{p} \geq 1 \text{ iff } \rho S \leq ((a - k)/2b)\ln((a - k)/2k).$$

(61) expresses the magnitude of $\bar{p}$ in terms of the parameters of the model. Suppose, for example, that $\rho = 0.02$ per year, $S/\bar{R} = 100$ years (i.e., at the current rate of extraction, the aquifer will be exhausted in 100 years), $(a - k)/2k = 20$ (e.g., $k = \$0.50$ and $(a - k) = \$20$). Then

$$\bar{p} = 20\exp(-2) \approx 7. \tag{62}$$

We should conclude that the value to be attributed to water at the margin is high (about 7 times extraction cost). As the date of exhaustion gets nearer, the accounting price rises to its upper bound, 20.

6. Exploration and Discoveries

How should one account for expenditure on explorations of new deposits of exhaustible resources? We imagine that the rate at which new reserves are discovered (N) is an increasing function of (1) current expenditure on explorations (E) and (2) the accumulated expenditure on explorations (M), but is a declining function of (3) accumulated extraction (Z). Denote the discovery function be $N(E_t, M_t, Z_t)$, where

$$dM_t/dt = E_t, \tag{63}$$
$$\text{and} \quad dZ_t/dt = R_t. \tag{64}$$

We revert to the model containing one manufactured capital good, K, and an exhaustible natural resource, S. In the familiar notation, $Y = F(K, R)$ is taken to be the aggregate production function. The remaining equations of motion are,

$$dK_t/dt = F(K_t, R_t) - C_t - E_t. \tag{65}$$
$$dS_t/dt = N(E_t, M_t, Z_t) - R_t. \tag{66}$$

The model has four capital assets K, S, M, and Z. Their accounting prices are denoted by p_K, p_S, p_M, and p_Z, respectively. Social welfare is given by (1). From Theorem 1, we have

$$dV_t/dt = \tag{67}$$
$$p_K[F(K_t, R_t) - C_t - E_t] + p_S[N(E_t, M_t, Z_t) - R_t] + p_M E_t + p_Z R_t.$$

There are two cases to consider:

(A) Assume that $\partial N/\partial M = 0$ (implying that $p_M = 0$) and $\partial N/\partial Z < 0$ (implying that $p_{Z_t} < 0$). Even in this case genuine investment is not the sum of investment in manufactured capital and changes in proven reserves ($N_t - R_t$). This is because new reserves are valued differently from existing reserves. Note too that exploration costs should not be regarded as investment.

Consider now the special case where the mining industry optimizes.[33] Then $p_K = p_S \partial N/\partial E$. If, in addition, $p_S N_t$ can be approximated by $p_K E_t$, one could exclude discoveries of new reserves from genuine investment, but regard instead exploration costs as part of that investment.

(B) Suppose $\partial N/\partial M > 0$. If the industry optimizes, we have

$$p_K = p_M + p_S \partial N/\partial E, \qquad (68)$$

and so $p_K > p_M$. It follows that genuine investment should now include not only new discoveries and investment in manufactured capital (as in Case A), but also exploration costs, using an accounting price that is less than that of manufactured capital.

7. Forests and Trees

As stocks, forests offer a multitude of services. Here we focus on forests as a source of timber. Hamilton and Clemens (1999) regard the accounting value of forest depletion to be the stumpage value (price minus logging costs) of the quantity of commercial timber and fuelwood harvested in excess of natural regeneration rates. This is an awkward move, since the authors do not say what is intended to happen to the land being deforested. For example, if the deforested land is converted into an urban sprawl, the new investment in the sprawl would be recorded in conventional accounting statistics.[34] But if it is intended to be transformed into farmland, matters would be different: the social worth of the land as a farm should be included as an addition to the economy's stock of capital assets. In what follows, we consider the simple case where the area is predicted to remain a forest.

Let the price of timber, in consumption numeraire, be unity and let ρ (assumed constant) be the social rate of discount. Holding all other assets constant, if B_t is aggregate forest land at, we may express social welfare as $V(B_t)$. The accounting price of forest land is then $\partial V_t/\partial B_t$, which we write as p_t.

Consider a unit of land capable of supporting a single tree and its possible successors. If the land is virgin, if a seed is planted at $t = 0$, if $F(T)$ is the timber yield of a tree aged T, and if T is the rotation cycle, then the present discounted value of the land as a tree-bearer is,

$$p_0 = F(T)e^{-\rho T}/(1 - e^{-\rho T}). \qquad (69)$$

Suppose instead that at $t = 0$ the piece of land in question houses a tree aged τ. What is the value of the land?

If the cycle is expected to be maintained, we have

$$p_0 = F(T)e^{-\rho(T-\tau)}/(1 - e^{-\rho(T-\tau)}). \qquad (70)$$

If instead the tree is logged now, but the cycle is expected to be maintained, the value of the land, after the tree has been felled, is given by (69). Depreciation of the forest, as a capital asset, is the difference between (70) and (69).

8. Human Capital

To develop an accounting framework for knowledge acquisition and skill formation, consider a modified version of the basic model of Section 2. In particular, the underlying resource allocation mechanism is assumed to be autonomous. Labour hours are assumed to be supplied inelastically and population is constant, we may as well then normalize by regarding the labour-hours supplied to be unity.

Production of the consumption good involves physical capital, K_{1t}, and human capital, H_{1t}. Here, H_{1t} is to be interpreted to be the human capital embodied in those who work in the sector producing the consumption good. Thus, if Y_t is output of the consumption good,

$$Y_t = F(K_{1t}, H_{1t}), \tag{71}$$

where F is an increasing function of its arguments.

Assume that human capital is produced with the help of physical capital, K_{2t}, and human capital, H_{2t}, and that, owing to mortality, it depreciates at a constant rate, γ. Output of human capital is given by the technology

$$G(K_{2t}, H_{2t}), \tag{72}$$

where G is an increasing function of its arguments and strictly concave, representing that the input of students is given.

By assumption, all individuals at a given moment of time have the same amount of human capital. Therefore, $H_{1t}/(H_{1t} + H_{2t})$ is the proportion of people employed in the sector producing the consumption good. Let the total stock of human capital be H. It follows that

$$H_{1t} + H_{2t} = H_t. \tag{73}$$

Write

$$K_{1t} + K_{2t} = K_t. \tag{74}$$

For simplicity of exposition, we assume that physical capital does not depreciate. Accumulation of physical capital can be expressed as

$$dK_t/dt = F(K_{1t}, H_{1t}) - C_t, \tag{75}$$

and the accumulation of human capital as

$$dH_t/dt = G(K_{2t}, H_{2t}) - \gamma H_t. \tag{76}$$

Since the resource allocation mechanism, α, is assumed to be autonomous, we have

$$V_t = V(\alpha, K_{1t}, K_{2t}, H_{1t}, H_{2t}). \tag{77}$$

Let p_{1t} and p_{2t} be the accounting prices of physical capital and q_{1t} and q_{2t} the accounting prices of human capital, in the two sectors, respectively (i.e., $p_{1t} = \partial V_t/\partial K_{1t}$, $q_{2t} = \partial V_t/\partial H_{2t}$, and so forth). Therefore, wealth can be expressed as,

$$Z_t = p_{1t}K_{1t} + p_{2t}K_{2t} + q_{1t}H_{1t} + q_{2t}H_{2t},$$

and genuine investment by

$$I_t = p_{1t}dK_{1t}/dt + p_{2t}dK_{2t}/dt + q_{1t}dH_{1t}/dt + q_{2t}dH_{2t}/dt. \tag{78}$$

Estimating q_{1t} and q_{2t} poses difficult problems in practice. It has been customary to identify human capital with education and to estimate its accounting price in terms of the market return on education (i.e., salaries over and above raw labour). But this supposes, as we have assumed in the above model, that education offers no direct utility. If education does offer direct utility (and it is widely acknowledged to do so), the market return on education is an underestimate of what we should ideally be after. Furthermore, human capital includes health, which too is both a durable consumption good and capital good.

An alternative is to use estimates of expenditures on health and education for the purpose in hand. Such a procedure may be be a reasonable approximation for poor societies, but it is in all probability far off the mark for rich societies.

If α were an optimum resource allocation mechanism, we would have $p_{1t} = p_{2t} = p_t$, say, and $q_{1t} = q_{2t} = q_t$, say. These prices would be related by the optimality conditions

$$U'(C_t) = p_t; \qquad p_t \partial F/\partial K_1 = q_t \partial G/\partial K_2;$$
and $\quad p_t \partial F/\partial H_1 = q_t \partial G/\partial H_2.$

9. Global Public Goods

Countries interact with one another not only through trade in international markets, but also via transnational externalities. Hamilton and Clemens (1999) include carbon dioxide in the atmosphere in their list of assets and regard the accounting price (a negative number) of a country's emission to be the amount it would be required to pay the rest of the world if carbon emissions were the outcome of a fully cooperative agreement. Their procedure is, consequently, valid only if each country is engaged in maximising global welfare, an unusual scenario. In what follows, we develop the required analysis.

Let G_t be the stock of a global common at t. We imagine that g is measured in terms of a "quality" index which, to fix ideas, we shall regard as carbon dioxide concentration in the atmosphere. Being a global common, G is an argument in the value function V of every country. For simplicity of notation, we assume that there is a single private capital good. Let K_{jt} be the stock of the private asset owned by citizens of country j and let α_j be j's (autonomous) resource allocation mechanism and α the vector of resource allocation mechanisms. If V_j is j's value function, we have

$$V_{jt} = V_j(\alpha, K_{jt}, G_t). \tag{79}$$

Let $p_{jt} = \partial V_{jt}/\partial K_{jt}$ and $g_{jt} = \partial V_{jt}/\partial G_t$. It may be that G is an economic "good" for some countries, while it is an economic "bad" for others. For the former, $g_j > 0$;

for the latter, $g_j < 0$. Let E_{kt} be the emission rate from country k and let γ be the rate at which carbon in the atmosphere is sequestered. It follows that

$$dG_t/dt = \sum_k E_{kt} - \gamma G_t. \tag{80}$$

Genuine investment in j is,

$$I_t = dV_{jt}/dt = p_{jt}dK_{jt}/dt + g_{jt}dG_t/dt,$$

which, on using (80), can be expressed as

$$I_t = p_{jt}dK_{jt}/dt + g_{jt}\left(\sum_k E_{kt} - \gamma G_t\right). \tag{81}$$

Notice that the expression on the RHS of (81) is the same whether or not α is based on international cooperation. On the other hand, dK_{jt}/dt and dG_t/dt *do* depend on how the international resource allocation mechanisms are arrived at (e.g., whether they are cooperative or non-cooperative); and they affect the accounting prices, p_{jt} and g_{jt}.[35]

10. Exogenous Productivity Growth

To assume exogenous growth in total factor productivity (the residual) over the indefinite future is imprudent. It is hard to believe that serendipity, unbacked by R&D effort and investment, can be a continual source of productivity growth. Moreover, many environmental resources go unrecorded in growth accounting. If the use of natural capital in an economy has in fact been increasing, estimates of the residual could be presumed to be biased upward. On the other hand, if a poor country were able to make free use of the R&D successes of rich countries, it would enjoy a positive residual.

The residual can have short bursts in imperfect economies. Imagine that a government reduces economic inefficiencies by improving the enforcement of property rights, or reducing centralized regulations (import quotas, price controls, and so forth). We would expect the factors of production to find better uses. As factors realign in a more productive fashion, total factor productivity would increase.

In the opposite vein, the residual could become negative for a period. Increased government corruption could be a cause; the cause could also be civil strife, which destroys capital assets and damages a country's institutions. When institutions deteriorate, assets are used even more inefficiently than before and the residual declines. This would appear to have happened in sub-Saharan Africa during the past forty years (Collins and Bosworth 1996).

We now study sustainability in the context of two models of exogenous productivity growth.

10.1. LABOUR-AUGMENTING TECHNICAL PROGRESS

Consider an adaptation of the model explored in Section 3. Physical capital and a constant labour force together produce a non-deteriorating all purpose commodity. The economy enjoys labour augmenting technological progress at a constant rate n. If K is capital and A is knowledge, we have in the usual notation,

$$Y_t = F(K_t, A_t), \tag{82}$$

$$dK_t/dt = F(K_t, A_t) - C_t, \tag{83}$$

and $\quad dA_t/dt = nA_t. \tag{84}$

There are two capital goods, K and A. Let p_K and p_A, respectively, be their accounting prices in utility numeraire. The sustainability criterion is then $p_K dK_t/dt + p_A dA_t/dt \geq 0$, or, equivalently,

$$dK_t/dt + q_t dA_t/dt \geq 0, \text{ where } q_t \equiv p_A/p_K. \tag{85}$$

It is instructive to study the case where the resource allocation mechanism is optimal. The equations of motion for p_K and p_A are,

$$dp_K/dt = \delta p_K - p_K \partial F/\partial K, \tag{86}$$

and $\quad dp_A/dt = \delta p_A - p_K \partial F/\partial A - np_A. \tag{87}$

Using (85)–(87) yields,

$$dq_t/dt = (\partial F/\partial K - n)q_t - \partial F/\partial A. \tag{88}$$

Suppose F displays constant returns to scale. Define k = K/A and c = C/A. Write $f(k) \equiv F(k, 1)$. From (83) and (84) we have

$$dk_t/dt = f(k_t) - nk_t - c_t, \tag{89}$$

or $\quad dk_t/dt = (\partial F/\partial K)k_t + \partial F/\partial A - nk_t - c_t.$

Adding (88) and (89) yields

$$d(q_t + k_t)/dt = (\partial F/\partial K - n)(q_t + k_t) - c_t. \tag{90}$$

It is simple to confirm that q + k is the present value of future consumption (discounted at the rate $\partial F/\partial K$) divided by A (the current state of knowledge). It follows that the sustainability criterion at t (condition (85)), divided by A_t, is

$$dk_t/dt + n(k_t + q_t) \geq 0. \tag{91}$$

10.2. RESOURCE AUGMENTING TECHNICAL PROGRESS

Consider an alternative world, where output, Y, is a function of manufactured capital (K) and the flow of an exhaustible natural resource (R). Let $A_t R_t$ be the effective supply of the resource in production at t and S_t the resource stock at t. Then we may write,

$$Y_t = F(K_t, A_t R_t), \tag{92}$$
$$dK_t/dt = F(K_t, A_t R_t) - C_t, \tag{93}$$
$$dA_t/dt = n, \tag{94}$$
$$dS_t/dt = -R_t. \tag{95}$$

There are three state variables. But we can reduce the model to one with two state variables. Thus, write $Q_t \equiv A_t R_t$ and $X_t = A_t S_t$. Then (93) and (94) become,

$$dK_t/dt = F(K_t, Q_t) - C_t, \tag{96}$$
and $\quad dX_t/dt = nX_t - Q_t. \tag{97}$

This is equivalent to a renewable resource problem, and the steady state is the Green Golden Rule, with

$$nX = Q. \tag{98}$$

Let p_K and p_X be the accounting prices of K_t and X_t, respectively. Then the sustainability condition is,

$$p_K dK_t/dt + p_X dX_t/dt \geq 0. \tag{99}$$

It is instructive to study the case where the resource allocation mechanism is optimal. Suppose also that F displays constant returns to scale. Following the approach of the previous example, let $q_t = p_X/p_K$. Then it is easy to confirm that

$$(dq_t/dt)/q_t = \partial F/\partial K - n. \tag{100}$$

Moreover, the optimal use of the productivity adjusted natural resource, Q_t, is determined by the condition,

$$\partial F/\partial Q = q_t. \tag{101}$$

Along the optimal programme, the sustainability condition (99) is,

$$F(K_t, Q_t) - C_t + q_t(nX_t - Q_t) \geq 0, \tag{102}$$
or $\quad (\partial F/\partial K)K_t + (\partial F/\partial Q)Q_t - C_t + q_t(nX_t - Q_t) \geq 0, \tag{103}$
or $\quad (\partial F/\partial K)K_t - C_t + nq_t X_t \geq 0. \tag{104}$

Inequality (104) says that consumption must not exceed the sum of capital income and the sustainable yield.

11. Exhaustible Resources: The Exporting Economy

The export of natural resources at given world prices raises issues similar to those we have just encountered in our analysis of exogenous productivity change. The exogenous "drift" term, $\partial V_t/\partial t$, in equation (8) has to be estimated.

Assume that extraction is costless. Suppose that at time τ the world market price of an exhaustible resource is q_τ. If R_τ is the volume of export, revenue is $q_\tau R_\tau$.

Write $C_\tau = q_\tau R_\tau$. (105)

The country's export policy, being governed by the underlying α, can be expressed as $R(\tau, S_t, t)$ for $\tau \geq t$. From equation (105) it follows that

$$dC_\tau/dt = q_\tau dR_\tau/dt = (\partial C_\tau/\partial S_t)dS_t/dt + q_\tau \partial R_\tau/\partial t, \quad (106)$$

As before, we assume that social welfare at t is,

$$V_t = \int_t^\infty U(C_\tau)e^{-\rho(\tau-t)}d\tau. \quad (107)$$

Let p_t denote the resource's accounting price. Since the criterion for sustainable well-being is dV_t/dt, we differentiate both sides of equation (107) with respect to t to obtain,

$$dV_t/dt = \quad (108)$$
$$-U(C_t) + \rho V_t + \int_t^\infty U'(C_\tau)[(\partial C_\tau/\partial S_t)dS_t/dt + q_\tau \partial R_\tau/\partial t]e^{-\rho(\tau-t)}d\tau.$$

But

$$dS_t/dt = -R_t.$$

Therefore, equation (108) reduces to

$$dV_t/dt = -U(C_t) + \rho V_t + p_t dS_t/dt + \int_t^\infty U'(C_\tau)e^{-\rho(\tau-t)}(\partial C_\tau/\partial t)d\tau. \quad (109)$$

Define $\mu(\tau, t) = \partial C_\tau/\partial \tau + \partial C_\tau/\partial t$. (110)

$\mu(\tau, t)$ can be regarded as an index of the extent to which the resource allocation mechanism is non-autonomous. Using equations (105)–(107) and (110), equation (109) can be reexpressed as,

$$dV_t/dt = -U(C_t) + \rho V_t + p_t dS_t/dt + \int_t^\infty U'(C_\tau)e^{-\rho(\tau-t)}\mu(\tau, t)d\tau \quad (111)$$
$$- \int_t^\infty U'(C_\tau)e^{-\rho(\tau-t)}(\partial C_\tau/\partial \tau)d\tau.$$

On partially integrating the last term on the RHS of equation (111) and cancelling terms, we obtain,

$$dV_t/dt = p_t dS_t/dt + \int_t^\infty U'(C_\tau)e^{-\delta(\tau-t)}\mu(\tau, t)d\tau. \quad (112)$$

The integral on the RHS of (112) is the "drift" term. As (112) shows, the index of sustainable welfare is the algebraic sum of genuine investment and the drift term. We now proceed to obtain simple rules for estimating the index in the case of two special non-optimum resource allocation mechanisms.[36]

Suppose C is constant.[37] In this case,

$$\partial C_\tau/\partial \tau = \partial C_\tau/\partial t = 0,$$

and $\mu(\tau, t) = 0$ in (112) is zero, and genuine investment measures changes in social welfare.

Suppose instead R is constant. It follows that

$$\partial R_\tau/\partial \tau + \partial R_\tau/\partial t = 0, \tag{113}$$

and $$\mu(\tau, t) = R_\tau \partial q_\tau/\partial \tau = q_\tau R_\tau (\partial q_\tau/\partial \tau)/q_\tau. \tag{114}$$

Using (113) and (114), we may write,

$$\int_t^\infty U'(C_\tau)e^{-\delta(\tau-t)}\mu(\tau, t)d\tau = \bar{\mu}_t/\delta, \tag{115}$$

where $\bar{\mu}_t$ can be interpreted as the average capital gains on the world market, as viewed from time t. Formally, (112) can be re-written as,

$$dV_t/dt = p_t dS_t/dt + \bar{\mu}_t/\delta. \tag{116}$$

12. Defensive Expenditure

How should defensive expenditure toward pollution control appear in national accounts? Denote by Q_t the stock of defensive capital and X_t investment in its accumulation. Let P_t be the stock of pollutants and Y_t aggregate output. We may then write,

$$dP_t/dt = G(Y_t, Q_t) - \pi P_t, \tag{117}$$

where $G(Y_t, Q_t) \geq 0$, $\partial G/\partial Y > 0$ and $\partial G/\partial Q < 0$.

Moreover, if defensive capital depreciates at the rate γ (> 0), then

$$dQ_t/dt = X_t - \gamma Q_t. \tag{118}$$

In the usual notation, the accumulation equation is expressed as,

$$dK_t/dt = F(K_t) - C_t - X_t. \tag{119}$$

Denote by p_t the accounting price of K, m_t that of defensive capital, and r_t (< 0) the accounting price of the pollutant. Wealth can then be expressed as,

$$p_t K_t + m_t Q_t + r_t P_t,$$

and genuine investment at t as,

$$I_t = p_t dK_t/dt + m_t dQ_t/dt + r_t dP_t/dt. \tag{120}$$

Equation (120) says that defensive expenditure against pollution ought to be included in the estimation of genuine investment ($m_t dQ_t/dt$), but, then, so should changes in the quality of the environment be included ($r_t dP_t/dt$). To include the former, but not the latter, would be a mistake.

13. Population Change and Sustainable Development

How does demographic change affect the index of sustainable development? There are a number of conceptual problems inherent in the welfare economics of reproductive behaviour that still remain usettled. Such problems have typically been bypassed in growth accounting; instead, it has been customary there to regard changes in population to be exogenously given. We follow that practice here.[38]

We seek to determine how population change influences the drift term ($\partial V_t/\partial t$) on the RHS of equation (8). An equivalent way of casting the problem is to regard population as a capital asset. Once we do that, what could appear to be a non-autonomous model reduces to an autonomous one. To illustrate, we adopt a natural extension of Harsanyi (1955) by regarding social welfare to be the average utility of all who are ever born. We formalize this 'dynamic average utilitarianism' as follows:

Let N_t be population size at t and $n(N_t)$ the percentage rate of change of N_t.[39] For notational simplicity, we ignore intragenerational inequality and changes in the age composition of the population. Let c_t denote per capita consumption at t. If C_t is aggregate consumption, $c_t = C_t/N_t$. Assume as before that labour is supplied inelastically in each period. Current utility of the representative person is $U(c_t)$ and social wefare is,

$$V_t = \int_t^\infty N_\tau U(c_\tau) e^{-\delta(\tau-t)} d\tau / \int_t^\infty N_\tau e^{-\delta(\tau-t)} d\tau.^{40} \tag{121}$$

If V_t is to be well-defined, we need to suppose that there exists $\varepsilon > 0$, such that $(\delta - \varepsilon)t > \int_0^t n(N_\tau) d\tau$ for large enough t. Notice though that, once we are given the population forecast, the denominator in (121) is independent of the policies that could be chosen at t. This means that a policy deemed to be *optimal* if (121) were used as the criterion of choice would also be judged to be optimal if instead social welfare V_t were taken to be of the form,

$$V_t = \int_t^\infty N_\tau U(c_\tau) e^{-\delta(\tau-t)} d\tau. \tag{122}$$

But for assessing whether or not a pattern of development sustains V_t, it matters whether V_t is taken to be (121) or (122).

Let K_{it} denote the stock of the ith type of capital good and write $k_{it} = K_{it}/N_t$. We now express by $\mathbf{k}_t$ the vector of capital stocks per head. The state variables are therefore $\mathbf{k}_t$ and N_t. We take it that α is autonomous. Then equation (121) implies that

$$V_t = V(\mathbf{k}_t, N_t). \tag{123}$$

Let the numeraire be utility. Define $\nu_t = \partial V_t/\partial N_t$. It is the contribution of an additional person at t to social well-being. ν_t is the accounting price of a *person* (as distinct from the accounting price of a person's human capital). Note that ν_t can be negative, depending on initial conditions at t and on the resource allocation mechanism.

Let p_{it} denote the accounting price of k_{it}. Equation (123) then implies

$$dV_t/dt = \sum_i p_{it} dk_{it}/dt + \nu_t dN_t/dt. \tag{124}$$

The RHS of equation (124) is genuine investment, inclusive of the change in the size of the population. It generalizes equation (8). We conclude that Proposition 1 remains valid so long as wealth comparisons mean comparisons of wealth *per capita, adjusted for demographic changes*.

In Arrow, Dasgupta, and Mäler (2003), we have studied optimal economies in which the adjustment term ($\nu_t dN_t/dt$) is not negligible, but nevertheless can be estimated in a simple way. Dasgupta (2001b) identified a set of circumstances where the term vanishes even in an imperfect economy. Suppose (i) $n(N_t)$ is independent of N_t; (ii) all the production processes are linear; and (iii) $c_t = c(\mathbf{k}_t)$, meaning that under the resource allocation mechanism α, *per capita* consumption is not a function of population size. In such circumstances V_t is independent of N_t (i.e., $\nu_t = 0$) and, so, equation (124) reduces to

$$dV_t/dt = \sum_i p_{it} dk_{it}/dt. \tag{125}$$

This finding can be summarised as

Theorem 6
If (i) $n(N_t)$ is independent of N_t, (ii) all the production processes are linear, and (iii) $c_t = c(\mathbf{k}_t)$, then social welfare is sustained at a point in time if and only if the value of the changes in per capita capital assets at that instant is non-negative.

The conditions underlying Theorem 6 are overly strong. It is tempting nevertheless to regard the value of changes in the per capita stocks of capital assets as a first approximation of dV_t/dt and then to estimate correction terms that reflect departures from the conditions underlying the theorem. That investigation is left for future work.[41]

14. Uncertain Productivity

How does future uncertainty in the productivity of capital assets influence accounting prices? In order to study this question in the simplest possible way, we revert to the Ramsey-Solow model of Section 3 and assume that the productivity of the single asset is uncertain. Analytically it is easiest to imagine that the underlying stochastic process generates a return on investment that is independently and identically distributed (iid) in each period. For convenience we now suppose that time is discrete (t = 0, 1, 2, ...). In what follows we indicate that a variable is random by placing a tilde over it. Let us denote the uncertain productivity of investment at date t by $\tilde{\mu}_t$. We assume that $\tilde{\mu}_t$ is non-negative and that the distribution of $\tilde{\mu}_t$ is atomless.

Population is assumed to be a constant and aggregate saving is taken to be a constant proportion, s, of wealth, where $0 < s < 1$. At each t the size of the capital stock that has been inherited from the previous period is a known quantity. Consumption is a fixed proportion $(1 - s)$ of that inherited stock. Therefore, assuming that capital does not deteriorate, the discrete time, stochastic counterpart of the accumulation equation (19) is,

$$\tilde{K}_{t+1} = (K_t - C_t)\tilde{\mu}_t,$$

from which we conclude that

$$\tilde{K}_{t+1} = s\tilde{\mu}_t K_t, \qquad t \geq 0,$$

and thus,

$$\tilde{C}_\tau = (1-s)K_t[_t\Pi^{(\tau-1)}(s\tilde{\mu}_k)], \qquad \tau > t \geq 0. \tag{126}$$

Writing by U(C) the utility of consumption, we take it that social welfare (V) is the expected value of the sum of discounted utilities over time. Letting E denote the expectation operator, this means that

$$V_t = E[\sum_t^\infty U(\tilde{C}_\tau)\beta^{(\tau-t)}], \qquad \text{where } \beta \equiv 1/(1+\delta) \text{ and } \delta > 0. \tag{127}$$

Suppose utility is iso-elastic. Let η be the elasticity of marginal utility. We consider the empirically interesting case, $\eta > 1$. We write U as:

$$U(C) = C^{1-\eta}/(1-\eta), \qquad \text{where } \eta > 1. \tag{128}$$

In (128), U is bounded above, but is unbounded below.

Write $E(\tilde{\mu}_t^{(1-\eta)}) = E(\tilde{\mu}^{(1-\eta)})$. If V_t is to be well-defined, we must now suppose that

$$\beta s^{(1-\eta)}E(\tilde{\mu}^{(1-\eta)}) < 1. \tag{129}$$

Using (126) and (128), and noting that the series in (127) is absolutely convergent, we can rewrite (127) as

$$V_t = -(1-s)^{(1-\eta)}K_t^{(1-\eta)}/(\eta-1)[1-\beta s^{(1-\eta)}E(\tilde{\mu}^{(1-\eta)})],$$

and, so, deduce that the asset's accounting price is

$$p_t = \partial V_t/\partial K_t = (1-s)^{(1-\eta)}K_t^{-\eta}/[1-\beta s^{(1-\eta)}E(\tilde{\mu}^{(1-\eta)})]. \quad (130)$$

How would changes in the distribution of $\tilde{\mu}_\tau$ ($\tau \geq t$) affect p_t? To study this, imagine that $\log(\tilde{\mu}_t)$ is normally distributed with mean m and variance σ^2. Denote the mean of $\tilde{\mu}_t$ by $\bar{\mu}$. In that case, we know that

$$\bar{\mu} = \exp(m + \sigma^2/2), \quad (131)$$

$$E(\tilde{\mu}^{(1-\eta)}) = \bar{\mu}^{(1-\eta)} \exp(-\eta(1-\eta)\sigma^2/2), \quad (132)$$

and $\quad \text{var}(\tilde{\mu}) = \bar{\mu}^2[\exp(\sigma^2)-1]. \quad (133)$

From (130)–(133) we confirm that, holding var($\tilde{\mu}$) constant, $dp_t/d\bar{\mu} < 0$. To study the effect of an increase in var($\tilde{\mu}$) on p_t, while keeping $\bar{\mu}$ constant, we must allow σ to increase in such a way that $(m + \sigma^2/2)$ remains unchanged. It is now a simple matter to confirm that $\partial p_t/\partial(\sigma^2) > 0$. And so, we have

Theorem 7
Other things the same, (i) if the expected return on investment were to increase, the assets' accounting price would decrease, and (ii) if the underlying risk in the asset's productivity were to increase, so would its accounting price increase.

Part (i) of Theorem 7 says that an increase in the expected rate of return on investment would lead to a decrease in the asset's accounting price, other things the same. But Part (ii) is also consistent with intuition. From (128) we know that utility, while bounded above, is unbounded below. We would then expect V_t to be particularly sensitive to the downside risk in $\tilde{\mu}$. Part (ii) of Theorem 7 says that if the risk in $\tilde{\mu}$ were to increase, the asset (at the margin) would become more valuable – other things the same. The Theorem's message should be expected to be even stronger if the underlying transformation possibilities among goods and services were to display thresholds, or, more generally, ecological non-convexities of the kind that is present in the model of the shallow lake (Section 4).[42]

Of course, consumers could be expected to respond to an increase in the mean return on investment, or to an increase in uncertainty in the return. What would be their response? We cannot tell unless we model the economic environment in which various parties make their saving decisions. The simplest place to look is an environment where the saving rate is optimal. There, people's response to a change in risk is also optimal. Levhari and Srinivasan (1969) have shown that in the model economy being studied here, if U is homogeneous of degree $(1-\eta)$ in C, the optimal saving ratio (s^*) is the solution of the equation,

$$s^\eta = \beta E(\tilde{\mu}^{(1-\eta)}). \quad (134)$$

Let us continue to assume that $\eta > 1$. From (130) and (134) we conclude that if the saving rate is optimal, then, other things the same, an increase in the expected return on investment leads to a decline in the accounting price of capital (i.e., $dp_t/d\bar{\mu} < 0$), and an increase in the riskyness of return leads to an increase in the accounting price (i.e., $dp_t/d\sigma^2 > 0$).[43]

Accounting prices of capital assets (as opposed to their market prices) are rarely estimated; but when they *are*, the estimates are mostly made on the basis of economic models that eschew uncertainty. The general moral of our finding here is that such studies underestimate the social worth of those assets.

15. Concluding Remarks

In this paper we have explored the way welfare analysis can be conducted in imperfect economies. In Sections 2–3 it was confirmed that the same set of accounting prices should be used both for the evaluation of policy reforms (e.g., project evaluation) and for assessing whether the economic programme being pursued sustains intergenerational welfare. In Sections 5–14 we studied the properties of accounting prices of environmental natural resources under a variety of institutional arrangements. We showed that for a number of cases it is possible to derive simple formulae for accounting prices. It was found that under plausible values of the relevant parameters, accounting prices of goods and services can be substantially different from their market prices.

A large empirical literature in ecology and epidemiology offers evidence that ecological processes are driven by non-convex transformation possibilities.[44] We note here in passing that metabolic processes also involve non-convex functional relationships between nutrition intake and nutritional status.[45] It was confirmed that accounting prices can be used in non-convex environments (Section 4). Our hope is that the methods developed here will be of use not only in environmental and resource economics (our focus of concern here), but also in nutrition and epidemiological studies.

Acknowledgements

Research support (for KJA) was provided by the William and Flora Hewlett Foundation. Preliminary results of our research were presented at a Workshop on "Putting Theory to Work: The Measurement of Genuine Wealth", held at the Stanford Institute for Economic Policy Research during 25–26 April 2002. Earlier drafts of the article were prepared when the authors visited the Abdus Salam International Centre for Theoretical Physics, Trieste, in May 2002 and the first two authors visited the Beijer International Institute of Ecological Economics, Stockholm, in August 2002. We are most grateful to Geir Asheim and Matteo Marsili for many helpful discussions and for correcting two errors in an earlier draft and to a referee for comments.

Notes

1. Dasgupta and Mäler (2000), Dasgupta (2001a, b), and Section 2.
2. For references to the technical literature on sustainable development, see Pezzey and Toman (2002).
3. Dasgupta and Mäler (2000), Dasgupta (2001a, b), and Section 2.
4. Serageldin (1995) and Pearce, Hamilton and Atkinson (1996) were early explorations of the practicalities of estimating a nation's comprehensive wealth.
5. The material in Section 2 has been taken from Dasgupta and Mäler (2000) and Dasgupta (2001a, b).
6. In a companion paper (Arrow, Dasgupta and Mäler 2003) we have developed criteria for identifying sustainable development under changing population size in optimizing economies.
7. Differentiability everywhere is a strong assumption. For practical purposes, however, it would suffice to assume that V is differentiable in K_i almost everywhere. The latter would appear to be a reasonable assumption even when production possibilities (including ecological processes) are realistically non-convex. See Section 4 below. However, if the location of these points on the space of capital stocks is uncertain and the uncertainty a smooth probability distribution, the *expected value* of V_t would be continuous.
8. It is not our purpose to review the several ways in which sustainable development can be, and has been, defined. Pezzey (1992) contains an early, but thorough, classification.
9. For convenience we have defined sustainability only for a moment in time. One could insist on the infinitely more demanding requirement: $dV_t/dt \geq 0$ for all t. Readers can confirm that our results can be rephrased in the obvious manner to be in accordance with this stiffer condition.
10. One of us (KJA) has produced an example of an optimum economic programme displaying the latter feature.
11. Pearce and Atkinson (1993) noted this result for optimizing economies.
12. Theorem 4 is, of course, familiar for economies where the government maximises social welfare (see e.g., Arrow and Kurz 1970).
13. If the project has been designed efficiently, we would have:

    $$\Delta Y_t = (\partial F/\partial K)\partial K_t + (\partial F/\partial L)\Delta L_t + (\partial F/\partial R)\Delta R_t,$$

    where F is an aggregate production function (Y = F(K, L, R)). The analysis that follows in the text does not require the project to have been designed efficiently. As we are imagining that aggregate labour supply is fixed, ΔL_t used in the project would be the same amount of labour displaced from elsewhere.
14. Dasgupta, Marglin and Sen (1972) and Little and Mirrlees (1974), respectively, developed their accounts of social cost-benefit analysis with consumption and government income as numeraire. Which numeraire one chooses is, ultimately, not a matter of principle, but one of practical convenience.
15. Thus

    $$q_t = \int_t^\infty U'(C_\tau)\partial C_\tau/\partial R_\tau e^{-\delta(\tau-t)}d\tau.$$

    Notice that if manufactured capital were to depreciate at a constant rate, say γ, the social cost of borrowing capital would be $\lambda_t = \delta + \gamma - (dp_t/dt)/p_t$.
    Let $\hat{q}_t$ be the accounting price of the resource *in situ*. At a full-optimum, $p_t\partial F/\partial R_t = q_t = \hat{q}_t$, and $U'(C_t) = p_t$.
16. To prove (16) notice that, by definition, ρ_t satisfies the equation

    $$U'(C_t)\exp(-\delta t) = U'(C_0)\exp(-\int_0^t \rho_\tau d\tau).$$

    If we differentiate both sides of the above equation with respect to t, (16) follows.
17. Notice that in imperfect economies δ and η may be unobservable. See Section 2.2.

18. Person-specific factors (e.g., age, health status, gender) can be included in the welfare function. This is routinely done in applied economics.
19. See the interchange between Johnson (2001) and Dasgupta (2001c) on this. For a more detailed analysis of the connection between environmental and resource economics and the economics of poverty, see Dasgupta (1982, 1993, 2000, 2003).
20. Estimates of the elasticity of marginal utility obtained from consumer behaviour, or, alternatively, from consumer responses to questions, have typically been in the range 1.5–2.5. The evidence thus acquired does not of course reflect what we mean by η here, but it is close enough.
21. These are not fanciful figures. Per capita consumption in a number of countries in sub-Saharan Africa declined over the past three decades at as high a rate as 1 percent per year, implying that for small values of δ, the consumption rate of interest would have been negative.
22. As the economy has a single asset, Theorem 3 is trivially true.
23. A special case of formula (32) appears in Dasgupta, Marglin and Sen (1972). However, unlike our present work, the earlier publication did not provide a rigorous welfare economic theory for imperfect economies.
24. Solow (1974) and Hartwick (1977) are the key articles on this limiting case.
25. For the ecology of shallow lakes, see Scheffer (1997) and Carpenter, Ludwig and Brock (1999).
26. Note too that because the resource allocation mechanism is imperfect, $-U_C \neq \partial V/\partial S$ (see Section 4.2).
27. Although, for ease of exposition, we are using the same notation, the points S_1, S_2, and S_3 here are not the same as the points S_1, S_2, and S_3 in the previous sub-section.
28. To the best of our knowledge, Kurz (1968) was the first to note that if utility depends directly on capital stocks, the optimality conditions may possess multiple stationary points even in a convex world. Skiba (1978) showed that in non-convex economies the optimality conditions may possess multiple stationary points even if the utility function is independent of stocks. The model of Brock and Starrett (2003) combines the two features.
29. Brock and Starrett (2003) refer to $\bar{S}$ as a Skiba point, the reference being to Skiba (1978).
30. McKelvey (1980) has studied a special case of the model of diffusion developed below.
31. In the limit, as λ tends to infinity, β tends to infinity, implying that depletion is instantaneous.
32. See also Hartwick and Hageman (1993) for a fine discussion that links El Serafy's formula to Hicks' formulation of the concept of national income (Hicks 1942).
33. That the industry optimizes does not mean that the economy is following an optimum programme.
34. It should be noted though that the value of urban land would be more than just the new investment: there is a contribution to the value (which could be of either sign) arising from changes in population density – both in the newly developed property and in places of origin of those who migrate to the property.
35. Social cost-benefit analysis, as sketched in Section 2.4, would enable a country to estimate whether it ought to alter its emissions. Nordhaus and Yang (1996) have studied international carbon emissions as the outcome of a non-cooperative equilibrium game among nations.
36. Asheim (1996), Sefton and Weale (1996), Vincent, Panayotou and Hartwick (1997), Aronsson and Löfgren (1998), and Cairns (2002) have published related findings, but in the context of optimising economies.
37. In this case the resource will be exhausted in finite time. For notational simplicity, we continue to present matters as though the horizon is infinite.
38. For a discussion of such problems and possible resolutions to the paradoxes that normative population theory has given rise to, see Dasgupta (2001b).
39. If N_t is a logistic function, $n(N_t) = A(N^* - N_t)$, where A and N^* are positive constants.
40. See Dasgupta (2001b) for a justification of this form of intergenerational welfare.
41. In Dasgupta (2001b) Theorem 6 was invoked to assess whether the world's poorest regions have experienced sustainable development in the recent past.

42. The reader can confirm that if $0 < \eta < 1$ in (128), then $dp_t/d\bar{\mu} > 0$ and $dp_t/d\sigma^2 < 0$; and if $\eta = 1$ (i.e., $U(C) = \log C$), then $dp_t/d\bar{\mu} = dp_t/d\sigma^2 = 0$. See the following footnote for an intuitive explanation for these results.
43. The reader can confirm that if $0 < \eta < 1$ in (128) and (134), then $dp_t/d\bar{\mu} > 0$ and $dp_t/d\sigma^2 < 0$. To understand the result, note that if $0 < \eta < 1$, then U is unbounded above, but bounded below. $\eta = 1$ corresponds to the case where $U(C) = \log C$. In this case s* is independent of both $\bar{\mu}$ and σ^2, and so $dp_t/d\bar{\mu} = dp_t/d\sigma^2 = 0$. The opposite pulls arising from the unboundedness of U at both ends cancel each other. See Hahn (1970) for an intuitive explanation for the way η influences the relationship between σ and s*.
44. See for example, Murray (1993).
45. On this see Dasgupta (1993).

References

Aronsson, T. and K.-G. Löfgren (1998), 'Green Accounting in Imperfect Economies', *Environmental and Resource Economics* **11**, 273–287.

Arrow, K. J., P. Dasgupta and K.-G. Mäler (2003), 'The Genuine Saving Criterion and the Value of Population', *Economic Theory* **21**, 217–225.

Arrow, K. J. and M. Kurz (1970), *Public Investment, the Rate of Return and Optimal Fiscal Policy*. Baltimore: Johns Hopkins University Press.

Asheim, G. B. (1996), 'Capital Gains and Net National Product in Open Economies', *Journal of Public Economics* **59**, 419–434.

Brock, W. and D. Starrett (2003), 'Non-Convexities in Ecological Management Problems', *Environmental and Resource Economics*, this issue.

Cairns, R. D. (2002), 'Green Accounting Using Imperfect, Current prices', *Environment and Development Economics* **7**, 207–214.

Carpenter, S. R., D. Ludwig and W. A. Brock (1999), 'Management of Eutrophication for Lakes Subject to Potentially Irreversible Change', *Ecological Applications* **9**, 751–771.

Collins, S. and B. Bosworth (1996), 'Economic Growth in East Asia: Accumulation versus Assimilation', *Brookings Papers on Economic Activity* **2**, 135–191.

Dasgupta, P. (1982), *The Control of Resources*. Cambridge, MA: Harvard University Press.

Dasgupta, P. (1993), *An Inquiry into Well-Being and Destitution*. Oxford: Clarendon Press.

Dasgupta, P. (2000), 'Population, Resources, and Poverty: An Exploration of Reproductive and Environmental Externalities', *Population and Development Review* **26**(4), 643–689.

Dasgupta, P. (2001a), 'Valuing Objects and Evaluating Policies in Imperfect Economies', *Economic Journal* **111**(Conference Issue), 1–29.

Dasgupta, P. (2001b), *Human Well-Being and the Natural Environment*. Oxford: Oxford University Press.

Dasgupta, P. (2001c), 'On Population and Resources: Reply', *Population and Development Review* **26**(4), 748–754.

Dasgupta, P. (2003), 'World Poverty: Causes and Pathways', in B. Pleskovic and N. H. Stern, eds., *Annual Bank Conference on Development Economics 2003*. Washington, DC: World Bank, forthcoming, 2004.

Dasgupta, P. and G. Heal (1979), *Economic Theory and Exhaustible Resources*. Cambridge: Cambridge University Press.

Dasgupta, P. and K.-G. Mäler (2000), 'Net National Product, Wealth, and Social Well-Being', *Environment and Development Economics* **5**, 69–93.

Dasgupta, P., S. Marglin and A. Sen (1972), *Guidelines for Project Evaluation*. New York: United Nations.

El Serafy, S. (1989), 'The Proper Calculation of Income from Depletable Natural Resources', in Y. Ahmad, S. El Serafy and E. Lutz, eds., *Environmental Accounting for Sustainable Development*. Washington, DC: World Bank.

Hahn, F. H. (1970), 'Savings and Uncertainty', *Review of Economic Studies* **37**, 21–24.

Hamilton, K. and M. Clemens (1999), 'Genuine Savings Rates in Developing Countries', *World Bank Economic Review* **13**, 333–356.

Harsanyi, J. C. (1955), 'Cardinal Welfare, Individualistic Ethics and Interpersonal Comparisons of Utility', *Journal of Political Economy* **63**, 309–321.

Hartwick, J. (1977), 'Intergenerational Equity and the Investing of Rents from Exhaustible Resources', *American Economic Review* **66**, 972–974.

Hartwick, J. and A. Hageman (1993), 'Economic Depreciation of Mineral Stocks and the Contribution of El Sarafy', in E. Lutz, ed., *Toward Improved Accounting for the Environment*. Washington, DC: World Bank.

Hicks, J. R. (1942), 'Maintaining Capital Intact: A Further Suggestion', *Economica* **9**, 174–179.

IUCN (1980), *The World Conservation Strategy: Living Resource Conservation for Sustainable Development*. Geneva: International Union for the Conservation of Nature and Natural Resources.

Johnson, D. Gale (2001), 'On Population and Resources: A Comment', *Population and Development Review* **27**(4), 739–747.

Kurz, M. (1968), 'Optimal Economic Growth and Wealth Effects', *International Economic Review* **9**, 348–357.

Levhari, D. and T. N. Srinivasan (1969), 'Optimal Savings Under Uncertainty', *Review of Economic Studies* **36**, 153–163.

Lipset, S. M. (1959), 'Some Social Requisites of Democracy: Economic Development and Political Legitimacy', *American Political Science Review* **53**, 69–105.

Little, I. M. D. and J. A. Mirrlees (1968), *Manual of Industrial Project Analysis in Developing Countries: Social Cost Benefit Analysis*. Paris: OECD.

Little, I. M. D. and J. A. Mirrlees (1974), *Project Appraisal and Planning for Developing Countries*. London: Heinemann.

Lutz, E., ed. (1993), *Toward Improved Accounting for the Environment*. Washington, DC: World Bank.

McKelvey, R. (1980), 'Common Property and the Conservation of Natural Resources', in S. A. Levin, T. G. Hallam and L. J. Gross, eds., *Applied Mathematical Ecology, 18: Biomathematics*. Berlin: Springer Verlag.

Murray, J. D. (1993), *Mathematical Biology*. Berlin: Springer-Verlag.

Nordhaus, W. D. and Z. Yang (1996), 'A Regional Dynamic General-Equilibrium Model of Alternative Climate-Change Strategies', *American Economic Review* **86**, 741–765.

Pearce, D. and G. Atkinson (1993), 'Capital Theory and the Measurement of Sustainable Development: An Indicator of Weak Sustainability', *Ecological Economics* **8**, 103–108.

Pearce, D., K. Hamilton and G. Atkinson (1996), 'Measuring Sustainable Development: Progress on Indicators', *Environment and Development Economics* **1**, 85–101.

Pezzey, J. C. V. (1992), *Sustainable Development Concepts: An Economic Analysis*, World Bank Environment Paper No. 2. Washington, DC: World Bank,

Pezzey, J. C. V. and M. A. Toman (2002), 'Progress and Problems in the Economics of Sustainability', in T. Tietenberg and H. Folmer, eds., *The International Yearbook of Environmental and Resource Economics 2002/2003*. Cheltenham, UK: Edward Elgar.

Portney, P. R. and J. P. Weyant, eds. (1999), *Discounting and Intergenerational Equity*. Washington, DC: Resources for the Future.

Ramsey, F. P. (1928), 'A Mathematical Theory of Saving', *Economic Journal* **38**, 543–549.

Scheffer M. (1997), *The Ecology of Shallow Lakes*. New York: Chapman and Hall.

Sefton, J. and M. Weale (1996), 'The Net National Product and Exhaustible Resources: The Effects of Foreign Trade', *Journal of Public Economics* **61**, 21–48.

Serageldin, I. (1995), 'Are We Saving Enough for the Future?', in *Monitoring Environmental Progress*, Report on Work in Progress. Washington, DC: Environmentally Sustainable Development, World Bank.

Skiba, A. K. (1978), 'Optimal Growth with a Convex-Concave Production Function', *Econometrica* **46**, 527–540.

Solow, R. M. (1956), 'A Contribution to the Theory of Economic Growth', *Quarterly Journal of Economics* **70**, 65–94.

Solow, R. M. (1974), 'Intergenerational Equity and Exhaustible Resources', *Review of Economic Studies* **41**(Symposium Issue), 29–45.

Tinbergen, J. (1954), *Centralization and Decentralization in Economic Policy*. Amsterdam: North Holland.

Vincent, J. R., T. Panayotou and J. M. Hartwick (1997), 'Resource Depletion and Sustainability in Small Open Economies', *Journal of Environmental Economics and Management* **33**, 274–286.

World Commission (1987), *Our Common Future*. New York: Oxford University Press.

The Economics of Non-Market Goods and Resources

1. John Loomis and Gloria Helfand: *Environmental Policy Analysis for Decision Making*. 2001   ISBN 0-7923-6500-3
2. Linda Fernandez and Richard T. Carson (eds.): *Both Sides of the Border*. 2002
                                                                                ISBN 1-4020-7126-4
3. Patricia A. Champ, Kevin J. Boyle and Thomas C. Brown (eds.): *A Primer on Nonmarket Valuation*. 2003   ISBN 0-7923-6498-8; Pb: 1-4020-1445-7
4. Partha Dasgupta and Karl-Göran Mäler (eds.): *The Economics of Non-Convex Ecosystems*. 2004   ISBN 1-4020-1945-9; Pb: 1-4020-1864-9

KLUWER ACADEMIC PUBLISHERS – DORDRECHT / BOSTON / LONDON